Chapter 1: The Power of Signage and Wayfinding Design

Significance of Signage and Wayfinding Design

In today's fast-paced and busy world, we are constantly bombarded with information, images, and stimuli. As we navigate through our daily lives, it is important that we are able to easily and efficiently find our way and access the information we need. This is where signage and wayfinding design play a crucial role. Signage and wayfinding design are not just about directing people from point A to point B, but they also have the power to enhance the user experience and create a sense of place. From the streets of a bustling city to the hallways of a corporate building, well-designed signs and wayfinding systems can make a significant impact on how people perceive and interact with their surroundings.

Understanding Semiotics

At the heart of effective signage and wayfinding design lies the use of semiotics. Semiotics is the study of signs and symbols and their interpretation and meaning. From simple road signs to complex wayfinding systems in airports, semiotics plays a crucial role in how we understand and navigate our environment. semiotics can be divided into three parts - signs, signifiers, and signified. Signs are the physical representation of an idea, while the signifiers are the visual elements of the sign. Signified refers to the meaning behind the sign. Understanding the relationship between these elements is essential in creating efficient and meaningful signage and wayfinding systems.

Research Methodology

In order to create effective signage and wayfinding systems, it is essential to conduct thorough research. This can include understanding the target audience, analyzing the environment, and studying the cultural and societal context. Through research, designers can gain insights into the needs and preferences of the users, which will help in creating a design that is not only visually appealing but also functional and user-friendly. Research methods commonly used in signage and wayfinding design include surveys, focus groups, observational studies, and user testing. These methods help gather quantitative and qualitative data, which are crucial in making informed design decisions. Now that we have an understanding of the significance of signage and wayfinding design and the role of semiotics, let's dive into the history of this field and how it has evolved over time.

Chapter 2: Evolution of Signage and Wayfinding Design

Evolution of Signage and Wayfinding Design

Signage has been around since ancient times, serving as a means of communication and a way to guide people to their destinations. The earliest forms of wayfinding involved primitive symbols and hand-drawn maps that were used to navigate through unfamiliar territory. As civilizations advanced, so did the techniques and methods of signage and wayfinding design. One of the earliest and most significant examples of signage and wayfinding design can be seen in Ancient Egypt. The Egyptians used hieroglyphics and cartouches to communicate and direct travelers to various destinations. These symbols were not only functional, but also served as an artistic expression of their culture and beliefs. This is a perfect example of the fusion of art and wayfinding, which has remained an integral element in the evolution of signage and wayfinding design.

The Industrial Revolution played a crucial role in the development of signage and wayfinding design. With the emergence of large cities and the rise of industrialization, there was a need for street signs, traffic signals, and other directional signages. As more people began traveling and commuting, the demand for efficient and effective wayfinding systems grew. This led to the development of standardized signage and wayfinding techniques that are still used today.

Key Influences

The evolution of signage and wayfinding design was not solely influenced by the development of technology and infrastructure. Art, language, and culture have also played a significant role in shaping the design principles and aesthetics of signage and wayfinding. The advent of Modernism and the rise of graphic design as a profession in the early 20th century had a profound impact on the design of signage and wayfinding. Modernist principles, such as minimalism and functionality, were incorporated into design, leading to the creation of visually striking and easy-to-understand signs. This was a major shift from the elaborate and ornate designs of the past. Another key influence on signage and wayfinding design was the development of Semiotics - the study of signs and symbols - in the 20th century. This field of study focused on how people interpret and make meaning out of visual communication, which played an essential role in the design of effective signage and wayfinding systems. Semiotics helped designers understand the psychology behind design, resulting in the creation of more impactful and intuitive signs.

Impact on User Experience

The evolution of signage and wayfinding design has had a significant impact on the user experience. As the world became more complex and crowded, people needed clear and concise directions to navigate their way through it. Signage and wayfinding became a vital aspect of user experience, providing guidance and reducing stress and confusion. Today, as digital technology continues to advance, the impact of signage and wayfinding design on user experience is even more pronounced. With the integration of technology, signs have become interactive and, in many cases, personalized. This allows for a more immersive and immersive wayfinding experience, catering to the individual needs and preferences of each user. Furthermore, signage and wayfinding design have become an integral part of branding and place-making. Effective signage not only guides people to their destination but also creates a sense of identity and experience for a place. This not only enhances the user experience but also adds to the overall aesthetic and cultural value of a location.

In conclusion, the evolution of signage and wayfinding design has been a dynamic and continuous process, influenced by various factors such as art, technology, and human psychology. It has played a vital role in shaping the user experience and will continue to evolve and adapt to the changing needs and trends of society.

Chapter 3: The Power of Semiotics in Signage and Wayfinding Design

Semiotics, the study of signs and symbols and how they create meaning, plays a crucial role in the design of effective signage and wayfinding systems. From icons and symbols to language and culture, this chapter explores the fundamental principles of semiotics and how they can enhance the user experience through effective communication. By understanding the power of semiotics, designers can create signage and wayfinding systems that are not only visually appealing, but also intuitive and user-friendly.

Semiotic Principles

At its core, semiotics is all about understanding how signs and symbols create meaning. In signage and wayfinding design, this involves considering the relationship between the sign (the visual representation) and the signified (the concept or idea it represents). This relationship is known as the signifier and the signified, respectively. When designing signage and wayfinding systems, signifiers must be carefully chosen to ensure they accurately convey the desired message to the user. This can involve taking into account factors such as color, shape, size, and placement. For example, a red stop sign is a universal symbol for stopping, making it an effective signifier for a navigation system. Another key principle of semiotics is the notion of sign codes. These are systems of meaning that are shared within a particular culture or community. By understanding the different sign codes that exist, designers can create signage systems that are easily understood and navigated by their intended audience.

Icons and Symbols

Icons and symbols are essential components of signage and wayfinding design. They are visual representations that convey meaning without the use of words, making them essential for creating signage that transcends language barriers. When designing icons and symbols, it is crucial to consider their universality and simplicity. The best icons and symbols are those that can be easily understood and recognized by a wide range of people. This involves avoiding complex or culturally-specific symbols and instead opting for those that are universally recognized, such as arrows, letters, and numbers. Additionally, cultural connotations must be taken into account when designing icons and symbols. For example, some cultures may interpret certain colors or shapes differently, which can significantly impact the effectiveness of a symbol. By conducting thorough research and user testing, designers can ensure that their icons and symbols are culturally appropriate and resonate with their intended audience.

Language and Culture

Language and culture are intertwined with semiotics and play a significant role in how people interpret signs and symbols. When designing signage and wayfinding systems, it is essential to consider the linguistic and cultural diversity of the user base. In some cases, a single language may not be sufficient for communicating with all users. This can include situations where multiple languages are spoken in a particular area, or where there is a high population of international visitors. In these cases, designers must carefully consider the use of multiple languages in their signage and ensure that each language is given equal visibility and prominence. Cultural considerations are also crucial when it comes to language in signage and wayfinding design. Certain phrases or words may have different connotations or meanings in different cultures, which can cause confusion or even offense. By collaborating with local experts and conducting thorough research, designers can create language that is culturally appropriate and effective.

In conclusion, semiotics is a fundamental aspect of signage and wayfinding design, and by understanding its principles and the power of icons, symbols, language, and culture, designers can create systems that enhance the user experience and effectively communicate with a diverse audience. By carefully considering these elements, designers can ensure that their signage and wayfinding systems not only look beautiful but also serve their intended purpose of guiding and informing users.

Chapter 4: The Signage Design Process

Signage and wayfinding design is a critical aspect of creating a positive user experience. It requires a thoughtful and well-planned approach to ensure that the intended message is effectively communicated to the target audience. The design process can be broken down into three stages: planning, design, and implementation. Each stage is essential and requires detailed attention to produce a successful outcome. In this chapter, we will delve into each stage of the signage design process and discuss the key elements that contribute to its success.

Planning Stage

The planning stage is the foundation of the signage design process. It sets the direction for the entire project and ensures that all stakeholders are on the same page. This stage involves research, analysis, and developing a clear understanding of the project's objectives. It is also a crucial time to identify potential challenges and develop strategies to overcome them. The first step in the planning stage is to gather all relevant information about the project. This includes the intended message, target audience, budget, and timeline. It is essential to have a clear understanding of the purpose of the signage to ensure that the design aligns with the project's goals. This information will also help guide the design decisions and facilitate communication between the designer and client. Research is a vital component of the planning stage. It involves gathering information about the project's location, target audience, and other relevant factors. For example, a signage project in a hospital will require a different approach than one in a shopping mall. Understanding the demographics, culture, and unique characteristics of the location and audience will help create a design that resonates with them. Another essential aspect of the planning stage is conducting a site analysis. This involves physically visiting the location and assessing the site's conditions and restrictions. It helps identify potential challenges, such as architectural barriers and accessibility issues, that may impact the design. By conducting a site analysis, the designer can develop solutions that are both aesthetically pleasing and functional.

Design Stage

The design stage is where all the pieces of the puzzle come together. It is the most creative and exciting part of the signage design process. Armed with the information gathered in the planning stage, the designer can now begin to develop concepts and designs that will effectively communicate the intended message to the target audience. The first step in the design stage is to develop a concept for the signage. This involves brainstorming ideas and visualizing how the signage will look and function in its intended environment. The designer will create sketches and mock-ups to help visualize the design's placement and scale. This stage also involves selecting the appropriate design elements, such as typography, colors, and graphics, that will best convey the desired message.

Once the concept is finalized, the designer will prepare digital renderings of the design. This will help the client visualize the final product and make any necessary adjustments. It is also a critical time to ensure that the design aligns with the project's budget and timeline. If any changes need to be made, this is the time to do so before moving on to the implementation stage.

Implementation Stage

The implementation stage is where the design comes to life. It involves turning the conceptual design into a physical product. This stage requires close collaboration between the designer, manufacturers, and installers to ensure that the final product is of the highest quality. The first step in the implementation stage is creating technical drawings of the design. These drawings will serve as blueprints for the manufacturer and ensure that all specifications are met. It is essential to work closely with the manufacturer to ensure that the design can be reproduced accurately and within the project's budget. Once the manufacturing process is complete, it is time for installation. This stage involves careful planning and execution to ensure that the signage is correctly placed and positioned. It is essential to consider factors such as visibility, readability, and accessibility during the installation process. By working closely with installers, the designer can ensure that the final product meets the project's objectives and vision.

In conclusion, the signage design process is a multi-faceted and intricate process that requires attention to detail and a collaborative approach. The planning stage sets the foundation for the entire project, the design stage brings the concept to life, and the implementation stage ensures that the final product is functional and visually appealing. By carefully following this process, designers can create a successful signage and wayfinding design that enhances the user experience and effectively communicates the intended message.

Chapter 5: Signage Design - Enhancing User Experience through Semiotics

Color and Contrast

When it comes to signage design, one of the most important elements to consider is color. Colors have the power to evoke emotions and create a sense of brand identity. The choice of colors used in signage can greatly impact the overall user experience and determine the effectiveness of a design. Color psychology plays a significant role in the design process, as different colors can have different impacts on individuals. For example, red is associated with energy, passion, and excitement, while blue is often associated with calmness, trustworthiness, and professionalism. Knowing the psychological impact of different colors can help designers choose the right color palette to achieve a desired emotional response from users. In addition to the psychology of color, contrast is also a crucial factor in signage design. Contrast refers to the difference in color between elements in a design. It helps make the information on a sign more legible and easily distinguishable. A good contrast of colors can make a sign stand out and catch the attention of individuals passing by, ultimately enhancing the user experience.

When selecting colors for signage, it's essential to keep in mind the purpose of the sign and the environment in which it will be displayed. For outdoor signage, colors that are bold and vibrant can be effective in grabbing the attention of passersby. In contrast, for indoor signage, softer and more muted colors may be more appropriate in creating a calming and professional atmosphere.

Typography

The typography used in signage design is another critical element that can greatly impact the user experience. Typography refers to the style, appearance, and arrangement of written content. When choosing fonts for signage, it's important to consider legibility and readability. Legibility is the ability to recognize and distinguish individual letters and words, while readability is the overall ease of reading and understanding written content. Signage should have fonts that are easy to read from a distance while also conveying the intended message clearly and concisely. In addition to legibility and readability, typography can also contribute to the visual aesthetic of a sign. Different fonts can evoke different emotions and create a sense of brand identity. For example, a serif font can convey a sense of traditional and formal branding, while a script font can evoke a more elegant and sophisticated feeling. Designers should also pay attention to font size and spacing when creating signage. An appropriate font size and spacing can make the sign more visually appealing and easier to read for users, ultimately enhancing their experience.

Composition

The composition of a sign refers to the arrangement and placement of elements within the design. A well-composed sign not only looks visually pleasing but also conveys its intended message effectively. It's essential to carefully consider the composition of a sign to ensure it is easy to read and understand. One important aspect of composition is hierarchy. This involves using elements such as size, color, and contrast to prioritize information on a sign. For example, the most critical information should be the most prominent and eye-catching, while secondary information should be smaller and less visually commanding. Another factor to consider in signage composition is balance. Balance refers to the distribution of visual weight within a design. Signs should have a sense of balance to create a harmonious and organized appearance. This balance can be achieved by placing elements in a way that is visually appealing and creates a sense of flow. Designers should also consider the placement of elements in relation to each other and the environment the sign will be displayed in. This can greatly impact the overall user experience, as a well-placed sign can make it easier for users to find their way and gather information.

In the world of signage and wayfinding design, color, typography, and composition all work together to create an effective and aesthetically pleasing result. By understanding the role of these elements and how they can enhance the user experience, designers can create signage that is not only informative but also visually appealing and memorable.

Chapter 6: Strategic Placement, Visibility and Legibility, Hierarchy and Layout

Signage and wayfinding design play a crucial role in guiding and informing individuals in various environments. As designers, it is our responsibility to ensure that our signage is strategically placed, highly visible, and legible, while also considering the hierarchy and layout of the information. In this chapter, we will delve deeper into these important aspects and discuss how they contribute to enhancing the user experience through semiotics.

Strategic Placement

Strategic placement of signage involves understanding the environment and the needs of its users. It includes considering factors such as where people are most likely to look, how much time they have to process information, and potential distractions in the environment. By carefully analyzing these factors, we can ensure that our signage is placed in the most effective and efficient locations. When designing signage for indoor environments, it is essential to consider the flow of foot traffic. Placing signs at key decision points, such as intersections or entrances, can help guide individuals in the right direction. As humans, we tend to follow the paths of least resistance, so placing signage in line with this natural flow can also be beneficial. In outdoor environments, the positioning of signage is crucial, as it must compete with other elements such as buildings, trees, and other visual clutter. To overcome these challenges, designers often use elevated structures, such as poles or arches, to make signage more visible and provide added prominence.

Visibility and Legibility

The visibility and legibility of signage are essential in ensuring that the information is effectively conveyed to the user. These factors also contribute to the overall readability and understandability of the message. If a sign is not visible or legible, it defeats the purpose and can result in confusion and frustration for the user. To improve visibility, designers often use high-contrast colors, bold graphics, and proper lighting.

High-contrast colors, such as black and white, are easily distinguishable and can help grab the user's attention. Bold graphics and images can also assist in capturing the user's interest and conveying the message more effectively.

Legibility refers to the ease with which someone can read and understand the information. This is determined by the font type, size, and spacing used in the design. The font type should be easily readable, clear, and consistent with the overall aesthetic of the environment. The size of the font is also essential, as it should be large enough to be visible from a distance but not too large that it overpowers the sign's design. Adequate spacing between letters and words also contributes to legibility, as it helps individuals process the information more efficiently.

Hierarchy and Layout

Hierarchy and layout are crucial components of effective signage design, as they determine the order in which information is presented and how it is visually organized. Hierarchy refers to the importance and emphasis given to specific pieces of information. By using different font sizes, colors, and placement on the sign, designers can create a hierarchy that guides the user's attention and helps them prioritize information. Layout refers to how the elements of the sign are arranged on the page. A well-designed layout should be simple and intuitive, with the user's eye naturally following the information in a logical order. Designers must consider the flow of information and any potential distractions or competing elements in the environment. A clean and organized layout can make a significant difference in the overall effectiveness of the sign.

In conclusion, strategic placement, visibility and legibility, and hierarchy and layout are all critical factors in ensuring that signage and wayfinding design effectively guide and inform individuals. By carefully considering these elements, designers can create signage that enhances the user experience and contributes to an overall positive aesthetic in the environment. As we continue to explore the role of semiotics in signage and wayfinding design, it is essential to keep in mind these key aspects and how they contribute to creating a successful design.

Chapter 7: Signage Types

Signage and wayfinding design serve a crucial role in enhancing the user experience. In this chapter, we will explore the different types of signage that contribute to creating effective wayfinding solutions. From identification signs to directional signs, each serves a specific purpose in providing information and guiding users through a space. Let us delve into the world of signage and discover its various forms and functions.

Identification Signs

Identification signs are essential in helping users recognize and locate a specific place, object, or service within a space. They serve as a visual cue, allowing users to identify where they are and where they need to go. These signs can take different forms, such as logos, names, and symbols. Identification signs often include a combination of text and images, using distinct colors and fonts to create a unique identity. In some cases, these signs may also include a brief description or tagline to provide additional information about the place or service. For example, a hospital may use an identification sign with the institution's name, logo, and tagline, such as "Healing Lives Every Day." This not only helps users identify the hospital but also creates a sense of trust and reassurance.

Informational Signs

Informational signs provide users with valuable information about a particular place, activity, or service. They can range from general information such as rules and regulations to specific details, such as opening hours and contact information. The purpose of informational signs is to help users make informed decisions and understand their surroundings better. These signs can take various forms and designs, depending on their purpose and location. For example, an informational sign in a park may use natural and earthy colors and include images of flora and fauna to blend in with its surroundings. On the other hand, an informational sign in a mall may use bright and bold colors with more prominent fonts to catch the attention of busy shoppers.

Directional Signs

Directional signs are crucial in guiding users from one location to another within a space. They provide information on the best route to take and help users navigate through complex environments. Directional signs are especially helpful in large and busy spaces such as airports, hospitals, and shopping malls. These signs often use arrows, symbols, and written instructions to guide users. They may also include the distance and time it takes to reach a particular destination. For example, a directional sign in an airport may display "Gate 22 - 10 minutes walk." This not only gives users an idea of the direction and distance but also allows them to plan their time effectively.

The Importance of Appropriate Signage Placement

In addition to the design and type of signage, placement is another crucial factor in creating effective wayfinding solutions. The position of each sign must be carefully considered to ensure it is visible and easily accessible to users. Proper placement of signage can promote seamless navigation and prevent confusion and frustration. Identification signs should be placed in prominent locations that are easily recognizable and visible from a distance. Informational signs must be placed in relevant areas, such as near an elevator or restroom. Directional signs should be strategically placed at decision points, such as intersections or forks in the road.

Using a Combination of Signage Types

To create a comprehensive wayfinding system, designers often use a combination of different signage types. This allows for a more efficient and multi-layered approach to communicating information and guiding users. A combination of identification signs, informational signs, and directional signs can work together to create a seamless and intuitive navigation experience. For example, a museum may use identification signs with the institution's logo and name in its entrance. Inside the museum, informational signs may provide details about the artwork and displays, while directional signs strategically placed throughout the space guide visitors from one exhibit to another.

The Power of Visual Design Elements

In all types of signage, visual design elements play a crucial role in creating an effective and aesthetically pleasing solution. Colors, typography, and images must work together to convey the desired message while also complementing the overall design of the space. Using appropriate colors and typography can not only improve the visibility and legibility of the signage but also create a cohesive visual identity. Images and symbols can also add meaning and context to the signs, making them more memorable and impactful.

A Harmonious Blend of Form and Function

Signage and wayfinding design is a delicate balance between form and function. While the primary purpose of signage is to provide information and guide users, it is also a form of visual communication that contributes to the overall aesthetic of a space. Finding the perfect balance between these two elements is crucial in creating an effective and harmonious wayfinding system.

In conclusion, identification, informational, and directional signs are the three primary types of signage that work together to form a cohesive wayfinding system. Each type serves a specific purpose and must be carefully designed and placed to ensure a seamless and user-friendly experience. By using a combination of these signage types and incorporating visual design elements, designers can enhance the user experience and create a successful wayfinding solution.

Chapter 8: Universal Design: Creating Inclusive Signage

Signage and wayfinding design plays a crucial role in creating a welcoming and accessible environment for all individuals. Whether navigating through a busy airport or finding a restroom in a shopping mall, well-designed signage is essential for ensuring a positive user experience for people of all abilities. As a designer, it is our responsibility to create a universal design that caters to the needs of every individual and promotes inclusivity. In this chapter, we will explore the important principles of universal design, ADA guidelines, and inclusive design in signage and wayfinding.

Universal Design Principles

Universal design is a design approach that focuses on creating environments, products, and services that are accessible and usable by people of all ages, abilities, and backgrounds. It involves designing solutions that are intuitive, flexible, and easy to use, regardless of an individual's physical or cognitive limitations. When applied to signage and wayfinding design, universal design principles can significantly improve user experience and create a welcoming and inclusive environment. The principles of universal design include equity, flexibility, simplicity, perceptible information, tolerance for error, low physical effort, and size and space for approach and use. These principles encourage designers to think beyond the norms and consider the diverse needs of individuals. For example, designing signage with high contrast colors and using both written and pictorial symbols can make information more accessible to people with visual impairments.

ADA Guidelines

The Americans with Disabilities Act (ADA) is a civil rights law that prohibits discrimination against individuals with disabilities in all areas of public life, including employment, transportation, and public accommodations. This law also applies to signage and wayfinding design, ensuring that people with disabilities have equal access to information and navigation.

According to the ADA guidelines, signage must have raised characters and Braille to accommodate individuals with visual impairments. The characters must be in a sans-serif font and have a high contrast ratio for easy readability. Additionally, ADA guidelines specify the appropriate size and placement of signage to ensure they can be easily seen and read by wheelchair users.

Inclusive Design

Inclusive design goes beyond universal design and takes into consideration the specific needs and preferences of individuals. It involves understanding the diverse abilities and limitations of people and designing solutions that cater to a broader range of users. Inclusive design promotes empathy and seeks to involve diverse perspectives in the design process. When it comes to signage and wayfinding design, inclusive design means creating solutions that cater to a wide range of individuals, including those with sensory processing issues, mental health conditions, and cognitive disabilities. This could involve incorporating tactile elements, using clear and concise language, and considering cultural and linguistic diversity in the design. Inclusive design also involves engaging with users throughout the design process to gather feedback and make necessary adjustments. This collaborative approach helps to create user-centered solutions that truly meet the needs of all individuals.

In conclusion, creating inclusive signage and wayfinding design involves understanding and applying universal design principles, adhering to ADA guidelines, and embracing an inclusive design approach. As designers, it is our responsibility to create an accessible and welcoming environment for individuals of all abilities. By incorporating these principles and guidelines into our design process, we can create solutions that enhance the user experience for everyone and promote inclusivity.

Chapter 9: The Role of Technology in Signage Design

Signage and wayfinding design have come a long way from simple painted signs and maps. With advancements in technology, the possibilities for creating effective and engaging signage have expanded. In this chapter, we will explore the impact of digital signage, interactive signage, and virtual reality on the user experience and how they can be integrated into the design process.

Digital Signage

Digital signage refers to the use of electronic displays to convey information. It has become an increasingly popular choice for businesses, institutions, and public spaces because of its flexibility and ability to capture attention. The use of high-quality graphics and animations can make digital signage stand out and create a unique brand experience. One of the biggest advantages of digital signage is its ability to display dynamic and customizable content. Unlike static signs, digital signages can be easily updated with new information, promotions, or events. This makes them ideal for environments with constantly changing information such as airports, shopping malls, and event venues. Digital signage also allows for targeted messaging and audience segmentation. With the use of data and analytics, businesses can tailor their content to specific demographics, times of day, or even weather conditions. This not only enhances the user experience but also increases the effectiveness of the signage. However, with all the possibilities that digital signage offers, it is essential to remember the fundamentals of good signage design. The same principles of clear and concise messaging, legibility, and placement still apply. Digital signage should also be incorporated into the overall design scheme and not distract from the surrounding environment.

Interactive Signage

Interactive signage takes digital signage one step further by allowing users to engage and interact with the displays. This type of signage can range from simple touch screens to more advanced technologies such as gesture control and augmented reality. The use of interactive signage can greatly enhance the user experience by providing a level of engagement and personalization. For example, interactive maps in public spaces can allow users to search for specific locations and get customized directions. In retail spaces, interactive displays can provide product information, promotions, and even allow customers to make purchases directly from the screen.

The key to successful interactive signage is simplicity and ease of use. Users should be able to navigate the interface and interact with the display without any confusion or frustration. The design should also consider accessibility and cater to users with different abilities.

Virtual Reality

Virtual reality (VR) is an immersive experience that allows users to be transported to a different environment or reality. While it may seem like a far fetched idea for signage design, VR can offer unique opportunities for storytelling and branding. In the retail world, VR can be used to create virtual fitting rooms or showrooms, allowing customers to try on clothes or view products before making a purchase. In healthcare, VR can be used to simulate medical procedures or provide virtual therapy sessions. As technology continues to develop, the possibilities for VR in signage design will only expand.

However, as with any new technology, there are challenges that come with incorporating VR into signage design. The cost can be a barrier for smaller businesses, and VR displays may require more space and maintenance. It is also essential to ensure that the experience is seamless and does not cause discomfort or motion sickness for users.

Integrating Technology into Signage Design

While digital signage, interactive signage, and virtual reality offer exciting possibilities, they should not be used for the sake of technology alone. It is crucial to consider how these elements can enhance the user experience and support the overall messaging and branding. When incorporating technology into signage design, it is important to keep in mind the user's journey and how they will interact with the signage. The placement and size of the displays should be carefully considered to ensure they are not overwhelming or distracting. The content should also be designed with the user in mind – it should be engaging, relevant, and easily digestible. As with any design element, it is essential to test and gather feedback from users to ensure that the technology is enhancing the overall experience. Designers should also stay informed about new technologies and trends to continue pushing the boundaries and creating innovative signage solutions.

In conclusion, technology has opened up new possibilities for signage design and has the potential to greatly enhance the user experience. However, it is crucial to use these elements thoughtfully and strategically to ensure that they align with the overall design and effectively communicate with the target audience. With a user-centered approach and a deep understanding of the technology at hand, designers can create impactful and memorable signage experiences.

Chapter 10: A User-Centered Approach: Enhancing Signage and Wayfinding Design

Understanding User Needs

In the world of signage and wayfinding design, it is crucial to have a user-centered approach. This means that every design choice should be made with the needs of the user in mind. After all, the purpose of signage and wayfinding is to guide and assist people in navigating their surroundings, so it is essential to understand their needs and preferences. One of the first steps in creating a user-centered approach is to conduct thorough research and gather information about the target audience. This can include understanding their demographics, cultural backgrounds, and their familiarity with the environment. For example, a signage design for a busy airport will differ from a signage design for a quiet university campus. Additionally, understanding the specific needs of different user groups is crucial. This can include individuals with disabilities, non-English speakers, and senior citizens. By understanding these needs, designers can ensure that their signage is accessible and inclusive to all users.

User Testing

Once the research on user needs is completed, it is essential to conduct user testing throughout the design process. This involves gathering feedback from actual users and incorporating it into the design. User testing can be conducted in various ways, such as surveys, focus groups, and observation. One of the benefits of user testing is that it allows designers to identify potential usability issues and make necessary changes before the signage is implemented. It also provides valuable insights into how users interact with the signage and any challenges they may encounter. To ensure that user testing is effective, it is crucial to involve a diverse group of users. This can include people of different ages, abilities, and cultural backgrounds. By gathering feedback from a diverse group, designers can create signage that is relevant and user-friendly for all.

Iterative Design

Incorporating user testing and feedback into the design process leads to an iterative approach. This means that the design is continuously evaluated and refined based on user feedback until it meets their needs and expectations. An iterative design process not only allows for constant improvement but also helps to minimize the risk of usability issues and costly changes after implementation. By testing the signage with users at various stages, designers can identify any problems or misunderstandings and make necessary adjustments before the final design is implemented. It is also crucial to involve stakeholders in the iterative design process, including the client, architects, and facility managers. By incorporating their feedback, designers can ensure that the signage design aligns with the overall vision and goals of the project. Designers should also keep track of all the changes made during the iterative design process to maintain a clear record of the design decisions. This will help to avoid confusion and maintain consistency in the final design.

In conclusion, a user-centered approach is crucial in creating effective signage and wayfinding design. By understanding the needs of users, conducting user testing, and using an iterative design process, designers can ensure that their signage is accessible, inclusive, and user-friendly for all. This approach not only leads to better user experiences but also minimizes potential issues and costly changes after implementation. As the world becomes more diverse and inclusive, it is essential for designers to prioritize a user-centered approach in their signage and wayfinding designs.

Chapter 11: Multilingual Signage, Cultural Sensitivity, and Localization

Multilingual Signage

In today's globalized world, it is crucial for signage to be easily understood by people from different linguistic backgrounds. As such, incorporating multiple languages in signage design has become a common practice. This not only ensures clear communication but also promotes inclusivity. The use of multilingual signage has many benefits. Firstly, it allows for effective communication with people from different language backgrounds, helping to bridge language barriers and make spaces more accessible. This is particularly important in areas with high tourist or immigrant populations. Additionally, multilingual signage also adds to the aesthetic appeal of a space by incorporating diverse languages and typography. It adds a sense of cultural richness and can showcase the diversity of a particular community or organization.

When designing multilingual signage, it is important to consider the order in which languages are presented. The most commonly used language in a particular area should be given priority, followed by other relevant languages. The placement of languages should also be strategic, ensuring that they are easily noticeable and legible for all users.

Cultural Sensitivity

Signage and wayfinding design is not just about conveying information, it is also about understanding and respecting cultural differences. As designers, it is important to be aware of cultural sensitivities and avoid any potential misunderstandings or offending any particular group. One aspect of cultural sensitivity in signage design is the use of symbols and icons. These should be carefully chosen and researched to ensure they are universally understood and do not hold any negative connotations in certain cultures. For example, the "thumbs up" gesture may be seen as a positive sign in Western cultures but is considered offensive in some Middle Eastern and Asian cultures.

Another crucial aspect is the use of language and phrasing. Certain words or phrases may have different meanings or undertones in different languages, and it is important to be mindful of this when designing signage for a multicultural audience. This also applies to color choices, as different colors hold different meanings and cultural significance in various societies.

Localization

Localization is the process of adapting a product or service to meet the language, cultural and other specific requirements of a particular region or audience. In the context of signage and wayfinding design, this means tailoring the design to specific cultural and linguistic contexts. To effectively localize signage, designers must conduct thorough research and consult with individuals from the target audience. This will ensure that the signage is not only culturally sensitive but also accurately conveys the intended message. Localization also involves taking into consideration physical and environmental factors, such as climate and local materials. For example, in areas with extreme weather conditions, signage materials should be durable and weather-resistant. Moreover, localization also promotes sustainability in signage design. By using materials that are readily available and familiar to the local community, designers can reduce the environmental impact of signage production.

In conclusion, multilingual signage, cultural sensitivity, and localization are all essential aspects of signage and wayfinding design. By incorporating these elements into the design process, we can create inclusive and culturally aware spaces that enhance the user experience and showcase the diversity of our global society.

Chapter 12: Signage in Different Environments

Indoor Signage

Signage is not just about guiding people from one point to another, it is also an opportunity to create an artistic and welcoming environment. Indoor signage plays a crucial role in enhancing the user experience, whether it's in a retail store, office building, or educational institution. Utilizing a user-centered approach, indoor signage can be strategically placed to inform and direct individuals, while also adding an aesthetic appeal to the space. Indoor signage comes in a variety of forms such as wall graphics, directories, informational panels, and room identification signs. These signs serve as landmarks, providing a sense of orientation and helping individuals navigate through complex indoor spaces. They also help create a sense of unity and coherence within the environment. When it comes to indoor signage design, it's important to consider the purpose and function of the space. For example, a healthcare facility signage will differ from that of a retail store. Healthcare facilities require clear and concise signage for ease of navigation, while retail stores may use signage as a tool for branding and marketing. The use of color, typography, and visual design elements play a significant role in creating effective indoor signage.

Outdoor Signage

As individuals approach a building or public space, the first thing they encounter is outdoor signage. This signage not only guides individuals to their destination but also serves as an initial impression of the space. Outdoor signage needs to be visually appealing and communicative, as individuals may not be familiar with the environment. Outdoor signage includes building identification signs, directional signs, and parking signs. These signs should be easy to read, well-lit, and placed strategically for maximum visibility. They should also reflect the identity and purpose of the space, in terms of color, style, and design elements. One of the challenges in outdoor signage design is the ever-changing environmental conditions. Signs must be durable, weather-resistant, and easy to maintain. Utilizing sustainable materials in outdoor signage design not only reduces the environmental impact but also ensures the longevity of the signs.

Transportation Signage

For transportation systems such as airports, bus stations, and train stations, signage is crucial for efficient navigation and movement of people. This type of signage must be clear, informative, and easily understood by individuals of all ages and languages. Transportation signage includes wayfinding signs, information kiosks, and navigation systems. These signs need to be strategically placed, taking into consideration the flow of traffic and user behavior. Color and typography play a vital role in transportation signage as they need to be easily identifiable and visible from a distance. As technology continues to advance, transportation signage is also evolving. Digital signage and interactive wayfinding systems are becoming increasingly popular, providing real-time information to users. These systems are not only informative but also add a modern and sophisticated touch to the transportation environment.

In addition to enhancing the user experience, transportation signage also plays a crucial role in safety and security. Emergency and evacuation signs must be clearly visible and easy to understand in case of any crisis situations.

In Conclusion

Indoor, outdoor, and transportation signage all contribute to the overall user experience and should be approached with careful consideration and attention to detail. By incorporating a user-centered approach, utilizing strategic placement and incorporating visual design elements, signage in different environments can enhance the aesthetic appeal and functionality of a space.

Chapter 13: Signage for Health and Safety: Enhancing the User Experience through Semiotics

When it comes to signage design, one of the most crucial areas is health and safety. In any environment, whether it be a workplace, school, or public space, the proper signage must be in place to ensure the well-being of individuals. The use of semiotics in designing these signs can greatly enhance the user experience, making them more effective and visually appealing. In this chapter, we will discuss the importance of emergency evacuation signage, health and safety regulations, and compliance in signage design.

Emergency Evacuation Signage

In the event of an emergency, clear and effective emergency evacuation signage can make all the difference. These signs must be strategically placed throughout a building or space to guide individuals to the nearest exit. But beyond just pointing the way, these signs must also convey a sense of urgency and seriousness. This is where the use of semiotics can be especially useful. The color red, for example, is often associated with danger and urgency. By incorporating this color into emergency evacuation signs, it immediately catches the eye and conveys the message that there is a potentially dangerous situation. Additionally, the use of arrows and symbols can also aid in quickly directing individuals to the nearest exit. Visual cues like these can help people navigate the space quickly and efficiently during a time of crisis.

Health and Safety Regulations

In many environments, there are specific health and safety regulations that must be followed. These regulations can cover a range of topics, from first aid procedures to proper disposal of hazardous materials. When designing signage for these regulations, there are a few key elements to consider.

Firstly, the language used must be simple and easy to understand. Not all individuals may be familiar with technical or medical terms, so signage should use clear and concise language. Additionally, the use of visuals can be beneficial in conveying important information to those who may have language barriers or difficulty reading. These visuals should be easy to interpret and relevant to the specific regulation being communicated.

Compliance

In addition to following regulations, it is also important for organizations to ensure that their signage is compliant with local laws and standards. This includes factors such as size, placement, and font type. Non-compliant signage can not only be dangerous in emergency situations, but it can also result in legal implications for the organization.

Semiotics can play a role in compliance as well. Different cultures may have specific symbols or colors that are considered taboo or offensive. It is important for designers to be aware of these cultural sensitivities when creating signage that will be used in diverse environments. This not only promotes inclusivity and respect but also avoids potential legal issues.

In Conclusion

Signage for health and safety is crucial in any environment, and effective design can greatly enhance the user experience. By incorporating semiotics, designers can create visually appealing and informative signage that not only conveys important information but also evokes a sense of urgency and compliance. From emergency evacuation signs to health and safety regulations, the use of semiotics can make a significant impact on the overall effectiveness of these signs.

Chapter 14: The Role of Branding in Signage and Wayfinding Design

In today's world, branding plays a crucial role in the success of any business or organization. It is not just about having a pretty logo, but rather the entire experience and perception that your brand evokes in the minds of your audience. And when it comes to signage and wayfinding design, branding becomes even more significant.

Importance of Branding

Branding is not just limited to the visual elements of a company, but it also includes the values, personality, and voice of the brand. It is what differentiates one company from another and helps create emotional connections with customers. In the context of signage and wayfinding design, branding is essential as it allows businesses and organizations to communicate their identity and messages visually. Signage is often the first point of contact between a business and its customers. It is like a first impression that can make or break a relationship. A well-designed and strategically placed signage that reflects the brand's identity can leave a lasting impression and attract potential customers. On the other hand, poorly designed or inconsistent signage can create confusion and even repel customers.

Consistency in Signage Design

Consistency is the key to successful branding. When all the signage elements, such as the font, color, graphics, and messaging, are consistent, it creates a visual harmony that helps the audience identify and connect with the brand. Consistency also adds a professional and cohesive look to the overall signage design and makes it easier for customers to navigate and find their way. Moreover, consistency in signage design also helps in building trust with customers. If a company's website and social media profiles show one brand identity, but their signage design is completely different, it can create doubts in the minds of customers about the authenticity of the brand. This is why it is crucial to maintain consistency in signage design to establish credibility and trust with customers.

Brand Identity

Brand identity is how a company wants to be perceived by its target audience. It includes visual elements such as the logo, color scheme, typography, and design elements that represent the brand. In signage and wayfinding design, it is essential to incorporate the brand's identity seamlessly to create a unified visual experience.

When designing signage and wayfinding systems, designers must consider the brand's identity and how it can be effectively represented through visual elements. The color palette, typography, and graphics used in the signage should align with the brand's identity and evoke the desired emotions and perceptions in the minds of the audience.

The Impact of Branding in Signage Design

Incorporating branding elements in signage and wayfinding design not only creates a visual connection with customers but also helps in reinforcing the brand's message and values. Signage can be used as an effective tool to promote a brand's unique selling proposition and create a memorable experience for customers. Moreover, when branding elements are consistently used in signage design, it helps in reinforcing brand recognition. Customers will start to associate certain colors, fonts, and graphics with the brand, making it easier for them to identify and remember the company.

Incorporating Branding in Signage Design: Best Practices

When incorporating branding in signage design, it is essential to keep the following best practices in mind:

- Use consistent colors, fonts, and graphics that align with the brand's identity
- Consider the location and surrounding environment when designing signage to ensure it stands out and is easily visible
- Include messaging that reflects the brand's voice and values
- Ensure the signage is easy to read and understand
- Regularly review and update signage to maintain a fresh and consistent look

Conclusion

In conclusion, branding plays a critical role in signage and wayfinding design. It allows businesses and organizations to create a visual identity that is consistent and memorable, thereby improving brand recognition and building trust with customers. By incorporating branding elements in signage design, companies can enhance their overall brand experience and leave a lasting impression on their audience.

Chapter 15: Signage and Wayfinding Design - Enhancing the User Experience through Semiotics

Cognitive Mapping

Cognitive mapping refers to the mental process of creating a visual representation of one's surroundings and the relationships between different locations. In the context of wayfinding, it plays a key role in understanding how people navigate and make sense of their environments. An effective signage and wayfinding design should aid in this process, making it easier for individuals to create a mental map and successfully navigate their way. A key way to facilitate cognitive mapping is through the use of consistent and intuitive symbols and graphics in signage. These symbols should be easy to understand and follow a universal design approach, especially when designing for a diverse range of users. This helps in reducing cognitive load and creating a more fluid navigation experience.

In addition, incorporating landmarks into signage design can also play a crucial role in cognitive mapping. These landmarks can serve as visual cues or reference points, helping individuals to orient themselves and make sense of their surroundings. They can range from physical objects such as buildings or natural features, to architectural or artistic features incorporated into the design of the signage itself.

Landmarks and Beacons

Landmarks and beacons are both visual cues that aid in cognitive mapping and wayfinding. Landmarks are fixed visual references that can be seen from a distance, while beacons are temporary visual cues that are used to guide users towards their destination. Both are essential in creating a sense of direction and helping individuals find their way. When incorporating landmarks and beacons into signage design, it is important to consider their visibility and impact on the surrounding environment. Ideally, they should be distinctive, easily recognizable, and integrated seamlessly into the design of the signage. This not only aids in navigation but also adds to the aesthetic appeal of the space.

In addition, incorporating culturally significant landmarks or beacons can add another layer of meaning to the signage design. This can help create a stronger connection between the user and the environment, making the wayfinding experience more personal and memorable.

Path Integration

Path integration is the process of estimating one's location and orientation within an environment based on one's movement and changes in surroundings. In wayfinding, this plays a crucial role in helping individuals navigate without getting lost. Effective signage and wayfinding design should take into account this process and provide clear and logical pathways for users to follow. One way to achieve this is through the use of directional cues in signage design. This can include arrows, maps, or other symbols that provide a sense of direction and guide individuals towards their intended destination. These cues should be strategically placed along the path to help users confirm their location and direction, reducing the risk of getting lost. Moreover, incorporating color and contrast can also aid in path integration. Bright and contrasting colors can be used to highlight important areas or paths, making them stand out and easier to follow. This can also add a sense of fun and playfulness to the signage design, making the wayfinding experience more enjoyable for users.

In conclusion, an effective signage and wayfinding design should focus on enhancing cognitive mapping, incorporating landmarks and beacons, and facilitating path integration for a seamless navigation experience. By considering the needs and abilities of a diverse range of users and incorporating elements of art and culture, designers can create a more engaging and inclusive wayfinding experience.

Chapter 16: Navigation Systems

When it comes to navigating through our surroundings, technology has become our trusted guide. Thanks to advancements in GPS technology, we can now easily find our way to a new destination without the stress and uncertainty of getting lost. However, as design professionals, it is important for us to consider not only the technological advancements but also the visual aesthetics and user experience of navigation systems in signage and wayfinding design.

Maps and Diagrams

For centuries, maps and diagrams have been used as tools for navigation. From ancient civilizations creating maps on stone tablets to modern digital maps on our smartphones, these visual representations of our surroundings have evolved to become an essential part of our navigation experience. However, when incorporating maps and diagrams into signage and wayfinding design, it is crucial to strike a balance between functionality and aesthetics.

Navigation Apps

With the rise of smartphones, navigation apps have become the preferred choice for many when it comes to finding their way around. These apps not only provide turn-by-turn directions but also offer additional features such as real-time traffic updates and route optimization. In the world of signage and wayfinding design, incorporating or collaborating with navigation apps can greatly enhance the user experience. By integrating digital maps and GPS technology into signage, users can seamlessly transition from navigating on their phones to following physical signage, creating a more cohesive navigation experience.

GPS Technology

GPS technology has become an essential part of our daily lives, from navigation systems to tracking our fitness activities. In signage and wayfinding design, GPS technology has opened up a whole new world of possibilities. By incorporating GPS technology into signage, we can create dynamic and interactive navigation systems that respond to the user's location and movements. This not only provides more accurate directions but also adds an element of excitement and surprise to the navigation experience. In addition to traditional GPS technology, there has been a rise in using beacon technology in signage design. Beacons use Bluetooth technology to communicate with devices within their proximity, providing customized and real-time information based on a user's location. This technology can be utilized in signage design to provide personalized navigation instructions and enhance the overall user experience.

Simplifying the Navigation Experience

While technology has undoubtedly made navigation easier, it is essential to ensure that the design of navigation systems is not overwhelming for the user. Too many directions or an overload of information can confuse or frustrate users, defeating the purpose of navigation. Design professionals must focus on creating a balance between providing necessary information and keeping the design visually appealing and easy to follow.

Incorporating a clear and intuitive user interface, simple and concise directions, and user-friendly features can greatly improve the navigation experience for users. As designers, we must also consider the needs of users with disabilities, such as those with visual impairments, and ensure that navigation systems are accessible and inclusive for all.

Integrating Navigation Systems into Design

When incorporating navigation systems into signage design, it is important to consider the overall design aesthetic and how the navigation system will fit in with the surrounding environment. The design should be visually appealing and not stand out as an eyesore. It should seamlessly integrate with the overall design of the space while still being easily identifiable for users.

As GPS technology and navigation systems continue to evolve, so must our approach to incorporating them into signage and wayfinding design. By keeping the user experience and visual aesthetics at the forefront of our design process, we can continue to enhance the navigation experience and create beautiful, functional, and intuitive navigation systems.

Chapter 17: Signage for Special Populations

When designing signage, it is important to consider the needs of all users, including those with specific challenges, such as children, elderly individuals, and visually impaired individuals. These populations require additional accommodations to ensure they can navigate their environment effectively and safely. In this chapter, we will discuss the various considerations and strategies for designing signage for these special populations.

Children

Children are a unique population when it comes to signage and wayfinding. On one hand, they may not be able to read or understand complex visual symbols, yet on the other hand, they have a natural sense of curiosity and exploration. When designing signage for children, it is important to keep the following in mind:

- Use simple language and visual symbols: Children may not have a well-developed vocabulary or understanding of abstract symbols, so it is best to use familiar words and simple visuals in signage. This will help them understand what the sign is trying to convey.

- Use bright, bold colors: Children are naturally drawn to bright, colorful objects, so incorporating bold colors in signage can help attract their attention and make it more engaging for them.

- Make it interactive: Children love to learn through play, so why not make signage a fun and interactive experience? This could include using puzzles, games, or incorporating interactive technology like QR codes or augmented reality.

- Ensure signage is at eye level: Children are much shorter than adults, so it is important to make sure signage is at their eye level to make it more accessible and easier for them to read and understand.

Elderly

Aging can bring about physical and cognitive changes that can make it challenging for elderly individuals to navigate their environment. When designing signage for the elderly, consider the following:

- Use large, clear fonts: As people age, their eyesight may decline, making it difficult to read small or intricate text. Using larger, clear fonts can make it easier for them to read and understand signage.

- Use high-contrast colors: Similar to changes in eyesight, aging can also affect color perception. Therefore, using high-contrast colors in signage can make it easier for the elderly to distinguish between different visuals and text.

- Ensure good lighting: Adequate lighting is crucial for the elderly to read and understand signage effectively. Make sure all signs are well-lit and there are no areas with dim lighting or shadows.

- Avoid clutter and keep it simple: As we age, our brains can become overwhelmed with too much information. Keeping signage simple and avoiding clutter can make it easier for elderly individuals to process and understand the information being conveyed.

Visually Impaired

Individuals with visual impairments require specific accommodations in order to navigate their environment safely. When designing signage for the visually impaired, consider the following:

- Use tactile and Braille signage: For individuals who are blind or have low vision, using tactile and Braille signage is essential. This allows them to feel and read the information on the sign through touch.

- Use high-contrast colors: Similar to designing for the elderly, using high-contrast colors can make it easier for those with visual impairments to distinguish between visuals and text.

- Use clear, simple language: When using text in signage, it is important to keep it clear and concise. Avoid using jargon or complex language that may be difficult for individuals with visual impairments to understand.

- Incorporate auditory cues: Auditory cues can be helpful for individuals with visual impairments to navigate their environment. This could include using sound or voice prompts on directional signage or incorporating audio description for digital signage.

In conclusion, designing signage for special populations requires careful consideration and accommodations to ensure their safety and accessibility. By keeping their unique needs in mind and employing the strategies discussed in this chapter, we can create signage that caters to everyone's needs, enhancing the overall user experience.

Chapter 18: Signage for Public Spaces

Public spaces are vital components of a community's social and cultural fabric. They are places where people gather, interact, and experience a sense of belonging. As such, they require signage that is not only functional but also reflective of their unique atmosphere and purpose. Parks, museums, stadiums, and shopping malls all fall under this category, and each presents its own set of challenges and opportunities for signage and wayfinding design. In this chapter, we will explore how signage can enhance the user experience in these public spaces through semiotics.

Parks

Parks offer a refuge from the chaos of city life and provide a natural setting for people to relax, play, and connect with nature. As such, the signage in these spaces should reflect this sense of tranquility and harmony. One way to achieve this is through the use of natural materials and colors in the signage design. Earthy tones, such as shades of green, brown, and blue, can create a sense of peace and serenity while also blending seamlessly with the park's surroundings. In contrast, the use of bold and bright colors would feel out of place and disrupt the overall aesthetic of the space. Moreover, parks often have multiple paths, trails, and areas that can be confusing for visitors to navigate. With the right signage, these spaces can be easily identified, and wayfinding becomes a more natural and enjoyable experience. Signage can be strategically placed at key points, such as at entrances, intersections, and major attractions, to guide visitors along their intended path.

Museums

Museums are symbols of culture, knowledge, and creativity. Therefore, the signage used in these spaces should be equally as refined and sophisticated. Since museums often display a wide range of exhibits, the signage must be clear and concise to provide visitors with the necessary information without overwhelming them.

In addition to providing direction and information, museum signage can also serve as a visual representation of the museum's brand and values. This can be achieved through the use of consistent typography, branding elements, and color schemes throughout the museum. By doing so, the signage not only serves a functional purpose but also adds to the overall aesthetic and atmosphere of the museum.

Stadiums

Stadiums are designed for large gatherings and are typically used for sports, concerts, and other events. The signage in these spaces needs to be bold, eye-catching, and able to withstand large crowds. In addition to providing direction and information, signage in stadiums can also serve as a way to enhance the fan experience and create a sense of excitement and anticipation. One way to achieve this is through the use of digital signage. Digital displays can be used to showcase live event updates, show highlight reels, and display interactive games for fans to play. This not only serves as a wayfinding tool but also adds an element of entertainment to the overall experience.

Shopping Malls

Shopping malls are complex spaces with multiple stores, restaurants, and amenities. With so much going on, the signage in these spaces needs to be informative and visually appealing. One way to achieve this is through creative wayfinding strategies, such as using distinctive icons and symbols for different store categories or incorporating interactive maps and directories. This not only makes the user experience more enjoyable but also encourages visitors to spend more time in the mall, ultimately leading to increased sales. Moreover, as shopping malls are often crowded with people, the signage needs to be highly visible and durable. The use of large-scale displays, bright colors, and clear typography can make sure that the signage stands out and is easily readable from a distance.

In conclusion, signage plays a crucial role in enhancing the user experience in public spaces. It serves as a tool for navigation, information, and brand representation. By utilizing semiotics and incorporating creative design strategies, the signage in parks, museums, stadiums, and shopping malls can not only fulfill their functional purpose but also add to the overall atmosphere and aesthetic of these spaces.

Chapter 19: Signage for Educational Spaces

Schools, Universities, Libraries

Educational spaces present unique challenges when it comes to signage and wayfinding design. These spaces are intended to inspire learning, creativity, and exploration, and the signage needs to reflect this atmosphere. At the same time, it is essential for signage to be informative, intuitive, and accessible to all users. In this chapter, we will explore how signage design can enhance the user experience and contribute to the overall atmosphere in schools, universities, and libraries.

Schools

Signage in schools not only serves the practical purpose of guiding students and staff to their destinations, but it also plays a crucial role in creating a positive learning environment. Signs in schools must be designed with the intended audience in mind: children of different ages and abilities. The use of child-friendly typography, vibrant colors, and engaging visuals can capture the attention of younger students and make the experience of navigating their school fun and exciting. One of the biggest challenges in school signage is creating a clear and consistent system that is easy for students to understand. This is especially crucial in schools with a large number of classrooms and buildings. Signage should be placed at key decision points, such as at the entrance to a building, in hallways, and at intersections. Visual cues such as symbols and arrows can also be used to guide students in the right direction.

Moreover, signage in schools can also serve as a form of education itself. For example, signs promoting sustainability and environmental awareness can be integrated into the design, sending a positive message to students about the importance of taking care of the world around them. It is also an opportunity to teach students about visual communication and how effective design can convey information effectively.

Universities

With sprawling campuses and a diverse student population, universities can benefit greatly from well-designed signage. Signage in universities serves the dual purpose of guiding students to their destinations and promoting the unique identity of the institution. Universities often have specific branding guidelines that need to be considered in the design of signage, ensuring consistency and a cohesive visual identity. Wayfinding in universities can be challenging, with multiple buildings and departments spread out over a large area. Different departments may have their own signage systems, which can create confusion for students. Thus, it is crucial to establish an overarching wayfinding system that is consistent across the campus. This can be achieved through the use of color-coded signs, clear maps, and directional arrows.

Additionally, universities often have a large international student population, making it essential to consider multilingual signage. This not only facilitates communication with international students but also creates a welcoming and inclusive environment. Multilingual signage requires a thoughtful design approach, taking into consideration the different languages and cultural norms of the students.

Libraries

Libraries are spaces designed for quiet contemplation and study, and the signage should reflect this atmosphere. Large, bold signs with loud graphics may not be suitable for a library, and instead, signs should be designed with simplicity and clarity in mind. Typography is especially crucial in library signage, as it can greatly affect the reading experience and the overall ambiance of the space. One of the main goals of library signage is to help users locate books and resources quickly. Therefore, signs should be strategically placed at the entrance and throughout the library, directing users to different sections and resources. The use of symbols and colors can also aid in navigation for those with visual impairments or language barriers. Furthermore, libraries often serve as community spaces, and it is essential to consider the needs of all users in the design of signage. This includes those with disabilities, such as wheelchair users, and those with sensory impairments. Braille, tactile signage, and audio options can all be incorporated to make the library experience more inclusive for all users.

In conclusion, educational spaces require a thoughtful and intentional approach to signage and wayfinding design. The goal should be to create a clear and intuitive system that not only guides users to their destinations but also contributes to the overall atmosphere and identity of the space. By incorporating the needs and preferences of the diverse users in these environments, signage can enhance the learning experience and leave a positive impression on all who enter.

Chapter 20: Signage for Healthcare Facilities

Healthcare facilities, whether hospitals, clinics, or nursing homes, are some of the most challenging environments to navigate. With constant movement of patients, visitors, and staff, it can be overwhelming and disorienting to find your way around. This is where effective signage and wayfinding design come into play, providing clarity and ease of navigation for all who enter these spaces.

Hospitals

Hospitals are often large, complex structures with multiple buildings and floors. It is essential to have clear and concise signage to help patients and visitors find their way to their desired destination without any added stress or confusion. In addition to providing directions, hospital signage should also serve as a source of comfort and reassurance for patients and their families during a potentially stressful time. To enhance the user experience, signage in hospitals should follow principles of user-centered design. This means considering the needs and abilities of all potential users, including the elderly, children, and people with disabilities. Large, easy-to-read fonts and clear pictograms can aid in readability for all, while Braille and tactile elements can assist those with visual impairments. Furthermore, effective signage in hospitals should also prioritize safety and emergency response. Clear and visible exit signs, evacuation routes, and emergency contact information should be prominently displayed throughout the facility. In the case of a crisis, well-designed signage can make all the difference in directing people to safety.

Clinics

Similar to hospitals, clinics also have high traffic and need for clear signage. However, clinics may have a more specific focus and target group, such as specialized medical practices or maternity clinics. In these cases, signage can play an essential role in branding and creating a cohesive atmosphere for patients. Using color psychology, clinics can use different colors and materials in their signage design to reflect their branding and create a soothing and welcoming atmosphere for patients. For example, a pediatric clinic may incorporate bright colors and playful illustrations, while a maternity clinic may opt for softer tones and calming imagery.

In addition to branding, clinic signage can also play a significant role in privacy and confidentiality. Patient rooms and waiting areas should be clearly marked to maintain privacy, and directions to specific areas, such as labs or radiology, should be discreet to protect sensitive patient information.

Nursing Homes

Signage in nursing homes serves a dual purpose – providing directions for residents and visitors while also acting as a memory aid for elderly residents. With an aging population, nursing homes need to consider the cognitive abilities of their residents when designing signage. Simple and easily understandable signage can help residents navigate and remember their surroundings, improving their overall quality of life. Furthermore, nursing homes can use signage to promote a sense of community and belonging among residents. Signage in communal areas, such as dining rooms or activity rooms, can use personalized messages and photographs to create a more homely and familiar environment. Incorporating artwork and cultural elements in signage design can also enhance the overall aesthetic and cultural sensitivity of a nursing home. By incorporating local landmarks or cultural symbols, residents can feel more connected to their surroundings and maintain a sense of identity.

In conclusion, effective signage and wayfinding design in healthcare facilities play a crucial role in enhancing the user experience and creating a stress-free environment for patients and visitors. By considering the needs of all potential users, prioritizing safety and emergency response, and incorporating branding and cultural elements, healthcare facilities can create a positive and welcoming atmosphere for all who enter.

Chapter 21: Signage for Corporate Environments

Offices

When it comes to corporate office spaces, signage plays a crucial role in creating a welcoming and efficient environment for both employees and visitors. From directional signs guiding people to different departments or meeting rooms, to informational signs providing important information about building safety or company policies, well-designed and strategically placed signage can enhance the overall user experience. One of the key considerations for signage in offices is readability. In a busy and often fast-paced environment, people need to be able to quickly and easily understand and follow signs. This is where the choice of typography and font size becomes important. Fonts that are clear and legible, with appropriate sizing, should be chosen for maximum impact. In addition to readability, the design of the signage should also align with the company's branding and aesthetics. This not only creates a cohesive visual identity, but it also helps employees and visitors feel connected to the company's values and goals. Utilizing the company's brand colors and incorporating its logo in signage can add a touch of personality to an otherwise sterile office space.

Innovative and creative solutions in signage design can also help enhance the overall atmosphere of the office and improve employee morale. For example, incorporating motivational quotes or images into signage can inspire employees and create a positive work environment. Additionally, interactive digital signage can provide useful information in a more engaging and dynamic way.

Conference Centers

Conference centers often see a high volume of traffic and the use of signage is essential in ensuring attendees can navigate the space efficiently. From directional signs to room and session schedules, signage needs to be strategically placed and easily understood. One key feature in signage design for conference centers is consistency. Using the same design elements and colors throughout all signs helps create a seamless and unified experience for attendees. This not only adds to the professionalism of the event but also helps attendees feel at ease and confident in their navigation. Another important feature for conference center signage is flexibility. With multiple events happening at the same time, signage may need to be updated or changed quickly. Utilizing digital signage that can be easily updated can save time and effort in making physical changes to signs. In addition to directional and informational signs, conference centers can also use signage to add to the overall ambiance of the event. Customized signs with event branding or themes can create a memorable and engaging experience for attendees.

Hotels

Signage in hotels not only serves a functional purpose but also plays a role in creating a welcoming and luxurious atmosphere for guests. From the moment a guest walks through the doors, signage should guide them seamlessly through the different areas and amenities of the hotel. First impressions are crucial, and the exterior signage of a hotel can set the tone for a guest's experience. Clear and eye-catching signs directing people to the entrance, parking, and valet services can create a positive first impression and make guests feel welcomed. Indoor signage, such as wayfinding signs and room numbers, should also align with the hotel's branding and aesthetics. Elegant and stylish signage can add to the overall luxurious feel of the hotel and enhance the guest's experience. Hotels can also utilize technology in their signage design. For example, interactive digital signs can provide guests with personalized information about amenities, events, and recommended local attractions. This not only adds convenience for guests but also showcases the hotel's commitment to modern and innovative technologies.

In conclusion, signage in corporate environments such as offices, conference centers, and hotels, goes beyond just providing information. It is an essential design element that contributes to the overall user experience and reflects the company's branding and values. By incorporating readability, consistency, flexibility, and creativity in signage design, these corporate spaces can create a cohesive and memorable experience for employees, visitors, and guests.

Chapter 22: Aesthetics in Signage Design

Balance

In the world of design, balance is an essential element that can make or break a project. It refers to the distribution of visual weight in a composition, creating a sense of equilibrium and stability. When it comes to signage design, balance plays a crucial role in creating a visually appealing and functional piece. When we talk about balance in signage design, we are not just referring to the physical symmetry of the sign, but also the balance of information and design elements. A well-balanced sign will have a harmonious combination of text, visuals, and negative space, making it easy to read and understand. A balanced sign will also draw the attention of the viewer without overwhelming them.

Achieving balance in signage design requires careful consideration of elements such as size, spacing, and color. The placement of text and images should be thoughtfully balanced to create a sense of harmony and unity in the design.

Proportion

Proportion refers to the size and scale of elements within a design in relation to each other. It is an essential aspect of aesthetics in signage design as it helps create a sense of hierarchy and importance. Proportion also plays a role in the readability and impact of a sign. When designing signage, it is crucial to consider the proportion of elements such as text, images, and negative space. Text that is too small or too large can be challenging to read and may not have the desired impact. Images that are out of proportion can also be visually unappealing and may take away from the overall message of the sign. A good rule of thumb is to use the "golden ratio" when determining the proportion of elements in signage design. This ratio, also known as the divine proportion or golden mean, has been used in art and design for centuries and is believed to be pleasing to the eye.

Harmony

Harmony is the concept of creating a sense of unity and coherence in a design. It involves the intentional use of color, typography, and imagery to create a cohesive and visually pleasing composition. In signage design, harmony is crucial for effectively communicating the intended message. When choosing elements for a sign, it is essential to consider how they work together to create a sense of harmony. Colors, fonts, and images should complement each other and enhance the overall design. The use of too many conflicting elements can create a sense of chaos and confusion in a sign. Creating harmony in signage design requires a keen eye for detail, an understanding of design principles, and a clear understanding of the message to be conveyed. When done correctly, the result is a sign that is visually appealing and effectively conveys the intended message.

Summary

Aesthetics play a vital role in signage design, and balance, proportion, and harmony are key elements in creating a visually pleasing and functional sign. Achieving the right balance of elements, proper proportion, and a sense of harmony is essential for effectively communicating the intended message and enhancing the user's experience.

As a designer, it is crucial to consider these aspects when creating signage designs. By paying careful attention to balance, proportion, and harmony, you can create impactful and visually appealing signs that will enhance the user experience and leave a lasting impression.

Chapter 23: Signage Design and Environmental Factors

When creating effective signage and wayfinding systems, it is crucial to consider how they will be impacted by environmental factors. These factors can include lighting, materials, and weather resistance and can greatly affect the durability and visibility of a sign. In this chapter, we will explore the importance of considering these factors in signage design and offer practical tips and solutions for addressing them.

Lighting

Lighting plays a crucial role in the visibility and legibility of signage. Poor lighting can render a sign unreadable and diminish its effectiveness. Therefore, it is essential to consider lighting conditions when designing and installing signage. The first step in addressing lighting is to understand the level of illumination in the location where the sign will be placed. This can vary depending on whether the sign is located indoors or outdoors. In outdoor environments, natural lighting can change throughout the day. Therefore, it is essential to choose materials and colors that will be visible in different lighting conditions. Another factor to consider is artificial lighting sources. Signs placed in areas with strong artificial lighting may require bolder and more vibrant colors to ensure their visibility. In contrast, signs in dimly lit areas may benefit from reflective materials to enhance their visibility.

Using different lighting sources can also add a dynamic and aesthetically pleasing aspect to signage. LED lighting, for example, can create a glowing effect that draws attention to the sign. However, it is important to ensure that these lighting sources do not interfere with the overall readability of the sign.

Materials

When it comes to signage design, choosing the right materials is crucial for both functionality and aesthetics. Different materials can offer varying levels of durability and weather resistance, so it is essential to consider the environmental factors a sign will be exposed to when making material choices. For outdoor signage, it is essential to choose materials that can withstand harsh weather conditions. This can include extreme temperatures, high winds, and heavy rain. Materials such as aluminum, acrylic, and high-pressure laminates are popular choices for their durability and resistance to weather damage. Indoor signage may not need to be as weather resistant, but it is still essential to choose materials that can withstand day-to-day wear and tear. Lightweight materials such as foam board and PVC are commonly used for indoor signs, as they are cost-effective and can be easily customized. In addition to durability, the choice of material also plays a significant role in the aesthetics of signage. Metallic materials can add a sleek and modern look, while wood and natural materials can create a warm and welcoming feel. As with all design decisions, the choice of materials should align with the overall branding and purpose of the signage.

Weather Resistance

Weather conditions can greatly impact the lifespan and visibility of signage. It is crucial to consider the climate in which the sign will be placed and select materials and design features that can withstand its effects. For outdoor signage, it is essential to consider factors such as rain, wind, and sun exposure. If a sign is constantly subjected to heavy rain, the material chosen should be waterproof to prevent damage and fading. Similarly, if a sign is placed in an area with strong winds, it should be designed with sturdy materials and fasteners to prevent it from being blown away. Sun exposure can also be a significant factor in the longevity of signage. Materials that are not resistant to UV rays can fade and deteriorate over time. Therefore, it is important to choose materials with UV protection or consider a protective coating to prolong the lifespan of the signage. Regular maintenance and upkeep can also help to mitigate the effects of weather. Inspecting and repairing any damage caused by weather can prevent the need for costly replacements in the future. Additionally, it is essential to assess and replace weather-specific features, such as grommets or fasteners, regularly.

In conclusion, environmental factors should be carefully considered in the design and creation of signage. By understanding the impact of lighting, materials, and weather, designers can create effective and durable signs that enhance the user experience and withstand the test of time. Stay tuned for the next chapter, where we will explore sustainability in signage design. As the world continues to face environmental challenges, it has become increasingly important for industries to make a conscious effort towards sustainability. This includes the field of signage and wayfinding design.

In this chapter, we will explore the various ways in which designers can make their signage more eco-friendly, energy-efficient, and reduce their overall environmental impact.

Chapter 24: Sustainable Signage Design: A Step Towards a Greener Future

Eco-friendly Materials

When it comes to signage materials, there are numerous options available. However, not all of them are eco-friendly or sustainable. As designers, it is our responsibility to choose materials that have a minimal impact on the environment. This includes materials that are either biodegradable, recyclable, or made from sustainable resources. One of the best options for eco-friendly signage materials is wood. Not only is wood a renewable resource, but it is also biodegradable and can be sustainably harvested. Other materials such as bamboo, cork, and reclaimed materials like metal or plastic can also be used in a sustainable way. These materials not only reduce environmental impact but also add a unique and aesthetically pleasing element to the signage.

Energy-Efficient Lighting

Lighting plays a crucial role in signage design, not only for visibility but also for creating an impact. However, traditional lighting methods can be wasteful and harmful to the environment. As designers, we can make conscious choices to use energy-efficient lighting options in our signage.

LED lights are an excellent alternative to traditional lighting. They consume significantly less energy and have a longer lifespan, reducing the need for frequent replacements. Additionally, LED lights do not contain harmful chemicals, making them an eco-friendly option. Integrating solar-powered lighting systems in outdoor signage can also contribute to reducing energy consumption.

Environmental Impact

The production, installation, and maintenance of signage can have a significant impact on the environment. From the resources used in the manufacturing process to the energy consumed in installation and upkeep, every aspect of signage creation can leave a carbon footprint. To reduce this impact, designers can opt for eco-friendly manufacturing processes that utilize sustainable materials and reduce energy consumption. Choosing locally sourced materials and utilizing efficient production techniques are also ways to lessen environmental impact. The maintenance and disposal of signage also play a role in minimizing environmental impact. By choosing durable materials and implementing proper maintenance practices, designers can reduce the need for frequent replacements and ultimately decrease waste. When it comes to disposing of old signage, designers should consider recycling options and properly dispose of any hazardous materials.

Sustainable Signage Design in Action

Implementing sustainable practices in signage design is not just beneficial for the environment, but it also presents a unique design opportunity. By incorporating natural materials and eco-friendly elements, designers can create a one-of-a-kind look for their signage. This not only sets the signage apart but also adds to the overall aesthetic of the space. A great example of sustainable signage design in action is the Green Streamline signage project in Vancouver, Canada. The project employed eco-friendly practices such as using sustainable materials, energy-efficient LED lighting, and solar-powered lighting systems in outdoor signage. The result was a visually striking, sustainable signage system that added to the city's green initiatives.

Collaborating for a Greener Future

To truly make a significant impact on the environment, it is essential for designers to collaborate with signage manufacturers and suppliers. By working together, designers can ensure that the materials used in signage production are sustainable and that the production process is environmentally friendly. Additionally, collaboration can also help in implementing sustainable maintenance and disposal practices. By working closely with manufacturers, designers can find innovative ways to recycle and refurbish old signage, reducing waste and environmental impact.

In conclusion, as designers, we must make a conscious effort towards creating sustainable and eco-friendly signage. By choosing sustainable materials, utilizing energy-efficient lighting, and reducing our overall environmental impact, we can contribute towards a greener future. Let's make sustainability a priority in our signage design and strive to create a positive impact on the environment.

Chapter 25: Maintaining Signage and Wayfinding Design

Deterioration and Wear

Signage and wayfinding design are crucial components of any built environment. They serve as guides, navigators, and communicators, providing necessary information and enhancing the user experience. However, as with any physical element, signage and wayfinding designs are prone to deterioration and wear over time. Factors such as weather, vandalism, and general wear and tear can significantly impact the effectiveness and aesthetic of these designs. As designers and creators, it is essential to consider these potential challenges and plan for long-term maintenance to ensure the longevity and quality of our work. One of the biggest culprits of signage and wayfinding design deterioration is exposure to the elements. The harsh rays of the sun, heavy rain, extreme temperatures, and strong winds can all take a toll on these designs. Over time, colors may fade, materials may warp, and letters or graphics may become illegible. In high-traffic areas, signs may also experience wear and tear from constant touching, banging, or scraping. In the case of outdoor signage, exposure to pollution, dirt, and other environmental factors can further accelerate the deterioration process.

Cleaning and Repairs

Regular cleaning and prompt repairs are vital in maintaining the quality and functionality of signage and wayfinding designs. If left unchecked, minor damages can quickly escalate and become more expensive and time-consuming to fix. Therefore, it is crucial to establish a maintenance plan that includes regular inspections and cleanings. Depending on the location and type of signage, cleaning may involve using specialized tools and products to remove dirt, grime, and other substances that may have adhered to the surface. In the case of electronic or digital signage, the cleaning process may involve updating software and ensuring proper functioning of all components.

For repairs, it is essential to have spare parts and materials readily available to minimize downtime and ensure consistency in design and branding. If a sign is damaged or vandalized, prompt action should be taken to prevent it from affecting the overall appearance and message. This also applies to any necessary repairs due to natural wear and tear. Regularly scheduled inspections can help identify any potential issues before they escalate and become more challenging and costly to fix.

Long-term Maintenance

In addition to regular cleaning and repairs, long-term maintenance is crucial in ensuring the effectiveness and longevity of signage and wayfinding designs. This may involve periodic replacements or updates to keep up with changing needs, technology, and styles. As with any design, signage and wayfinding should also be periodically evaluated for relevance and effectiveness. This may include seeking feedback from users and conducting studies to track changes in behavior and understanding. When planning for long-term maintenance, it is essential to consider the sustainability of the design. Using durable and environmentally-friendly materials can help reduce the need for excessive repairs and replacements, minimizing waste and cost in the long run. It is also essential to consider the accessibility of signage and wayfinding designs, ensuring that they are inclusive and can be easily understood and used by all individuals.

In conclusion, while creating and implementing signage and wayfinding designs is a crucial step, maintaining them is equally important. By planning for regular cleaning, prompt repairs, and long-term maintenance, we can ensure that these designs continue to enhance the user experience and effectively fulfill their purpose for years to come. As designers, it is our responsibility to create not just aesthetically pleasing designs, but also sustainable and functional ones that can withstand the test of time.

Chapter 26: Budgeting and Cost Considerations for Signage and Wayfinding Design

When it comes to designing effective signage and wayfinding systems, it's important to consider not only the visual and functional elements, but also the financial aspect. As with any design project, budgeting and cost considerations play a crucial role in the decision-making process. In this chapter, we will delve into the various factors that can impact the overall cost of signage and wayfinding design, and provide insights on how to manage and budget for these expenses.

Budgeting

Creating a budget for signage and wayfinding design is essential in order to ensure that the project stays on track and within the client's financial capabilities. Before even beginning the design process, it's important to consult with the client and establish a clear understanding of their budget constraints. This will help determine the scope of the project and the available resources. One thing to keep in mind when budgeting for signage and wayfinding design is that costs can vary greatly depending on the project's size and complexity. A small, single-location project will naturally require a smaller budget compared to a large-scale project that spans multiple locations or requires custom elements. In addition to the overall budget, it's also important to allocate specific amounts for different aspects of the project, such as design, materials, and installation. This will help ensure that the design team stays within the allotted budget and avoids overspending.

Material Costs

The materials used in signage and wayfinding design can greatly impact the overall cost of the project. Therefore, it's important to carefully consider and select materials that not only fit within the budget, but also meet the functional and aesthetic needs of the project. One way to manage material costs is by incorporating cost-effective materials without sacrificing quality. For example, instead of using expensive metals for lettering, consider using lower-cost materials such as vinyl or acrylic with a metallic finish. Additionally, using durable and long-lasting materials can also help reduce maintenance and replacement costs in the long run.

Another important factor to keep in mind is the sustainability of the materials used. Choosing eco-friendly materials not only reduces the project's environmental impact, but can also potentially save costs in the long term.

Installation Costs

The installation process is often an overlooked factor when budgeting for signage and wayfinding design. However, it's important to consider these costs in order to avoid any surprises down the road. The complexity of the installation process can greatly impact the overall cost. For example, if a project involves designing and installing custom signage in a difficult-to-reach location, it may require additional equipment and personnel, thus increasing the installation costs. Collaboration and communication between the design team and the installation team is also key in managing installation costs. By working closely together, the team can ensure that the designs are feasible and the installation process is efficient, thus reducing any unnecessary expenses.

Managing Costs for a Successful Project

Managing and budgeting for costs is an important aspect of any signage and wayfinding design project. By collaborating closely with the client, selecting cost-effective and sustainable materials, and carefully considering installation costs, the project can stay on track and within budget. Another helpful tip in managing costs is to regularly review and adjust the budget as needed throughout the design process. This allows for any unexpected expenses or changes in scope to be addressed and managed effectively.

In addition, working with a trusted and experienced signage manufacturer can also help control costs and ensure a successful outcome. They can provide valuable insights and recommendations on materials and installation techniques, helping to keep costs in check.

Conclusion

In conclusion, when it comes to designing signage and wayfinding systems, it's essential to not only consider the aesthetic and functional aspects, but also the financial aspect. By setting a clear budget, carefully selecting materials, and managing installation costs, the project can stay within budget and deliver a successful, effective design that enhances the user experience.

Chapter 27: Signage Regulations and Compliance Challenges

When designing signage for a project, there are numerous factors to consider such as the target audience, location, message, and aesthetics. However, another crucial aspect that must not be overlooked is the regulatory requirements for signage. These regulations are in place to ensure that signage is safe, informative, and serves its purpose effectively. As signage designers, it is crucial to stay up-to-date with local, state, and federal regulations to avoid non-compliance issues.

Local Regulations

Local regulations for signage can vary widely from city to city and even within a city's different neighborhoods. This is why it is essential to thoroughly research and understand the specific signage regulations for the location of your project. Local regulations typically govern the size, height, and placement of signage. They may also have restrictions on the materials and lighting used, as well as the frequency of sign changes. It is important to work closely with the city's planning department to ensure that your signage design complies with their regulations.

State Regulations

In addition to local regulations, designers must also be aware of state signage laws. These may include requirements for the use of specific symbols or colors on certain signage, as well as specifications for signs in public places, such as restrooms and emergency exits. Some states may also have restrictions on the use of digital signage or require permits for certain types of signage. It is crucial to research and comply with these state regulations to ensure that your signage is legal and meets the necessary standards.

Federal Regulations

Federal regulations for signage focus mainly on safety and accessibility. The Americans with Disabilities Act (ADA) sets standards for signage that must be followed to ensure that people with disabilities have equal access to information. This includes specifications for the font size, contrast, and Braille on signage in public places. The Occupational Safety and Health Administration (OSHA) also has regulations for signage in workplaces, particularly related to health and safety. These federal laws must be carefully studied and followed to avoid any legal issues.

Compliance Challenges

While regulations are in place to ensure safety and accessibility, they may present challenges for designers. These regulations can limit creativity and may require adjustments to designs to meet specific font sizes, colors, or materials. Additionally, obtaining permits and complying with regulations can add extra time and costs to a project. However, it is crucial to remember that compliance with regulations is essential for the safety and well-being of the public. Another compliance challenge for designers is keeping up with the ever-changing regulations. Local and state laws may be revised, and new federal regulations may be implemented, requiring designers to update their knowledge and ensure that their designs comply with the latest requirements. This can be time-consuming and may even require redesigning of existing signage. In some cases, designers may also face challenges when attempting to comply with conflicting regulations. This can often occur when designing signage for a project that spans across different cities or states. In these instances, designers must carefully navigate and balance the requirements to ensure that the signage is compliant with all regulations without compromising its effectiveness.

In Conclusion

As signage designers, it is our responsibility to understand and comply with all local, state, and federal regulations. Failure to do so may result in legal consequences, but more importantly, it can compromise the safety, accessibility, and effectiveness of our signage. While compliance challenges may arise, it is crucial to prioritize the well-being of the public and ensure that our designs meet all regulatory standards.

Chapter 28: User-Centered Evaluation

The ultimate goal of signage and wayfinding design is to enhance the user experience and provide effective navigation. In order to achieve this, it is essential to continuously evaluate the effectiveness of the design from the perspective of the user. This chapter will delve into different aspects of user-centered evaluation in signage and wayfinding design, including navigation patterns, user interaction, and user feedback.

Navigation Patterns

One of the primary goals of signage and wayfinding design is to guide users efficiently from one place to another. To ensure the effectiveness of this navigation, it is crucial to understand and analyze different navigation patterns. These patterns can vary depending on the environment, purpose, and user demographics.

In urban environments, for example, wayfinding design must consider fast-paced movement and constant distractions. In such scenarios, clear and concise signage with bold visual cues is necessary to quickly orient and guide users. On the other hand, in healthcare facilities, wayfinding should prioritize ease of navigation for patients and their families, many of whom may be experiencing stress and anxiety. This may call for more detailed and empathetic signage design that takes into account the specific needs and mental state of the users.

User Interaction

Effective user-centered evaluation must also consider the interaction between the user and the signage design. This can include physical interaction, such as the ease of reading and comprehending the signs, as well as emotional interaction, such as the feeling of safety and comfort that the signage provides. This is where the semiotic framework plays a crucial role. Signage that is designed with semiotics in mind can communicate messages and information effectively, regardless of the user's background or language. The use of visual cues, familiar symbols, and clear messaging can enhance user interaction with signage, making it easier for users to navigate and understand their surroundings.

User Feedback

Gathering feedback from users is a vital component of user-centered evaluation. User feedback can provide valuable insights into the effectiveness of signage design and identify areas for improvement. This can be done through surveys, interviews, and even observation of user behavior. Moreover, involving users in the design process from the beginning can also lead to better user feedback. By understanding the needs and preferences of the users, designers can create signage that resonates with their target audience and effectively guides them.

Furthermore, in today's digital age, user feedback can also be obtained through social media and online platforms. This allows for continuous and real-time evaluation of the signage design, leading to quicker adaptation and improvement. User-centered evaluation is an essential aspect of signage and wayfinding design that cannot be overlooked. By understanding navigation patterns, prioritizing user interaction, and gathering user feedback, designers can create effective and impactful signage that enhances the user experience and promotes efficient navigation. With the constant evolution of technology, there are also more opportunities to gather and analyze user feedback, making it easier to continuously improve and refine signage design. By putting the user at the center of the design process, signage and wayfinding design can truly enhance the user experience and make navigation a seamless and enjoyable experience.

Chapter 29: Signage Regulations: Protecting Your Design

As a signage and wayfinding designer, you put countless hours and creative energy into crafting the perfect design for your client's space. It is important to not only create beautiful and functional signage, but also to ensure that your design is protected from any potential copyright, trademark, liability, and intellectual property issues. In this chapter, we will delve into the regulations and legal considerations that are crucial to understand as a signage designer.

Copyright and Trademark

One of the first things to consider when designing signage is copyright and trademark laws. Copyright protects the original expression of ideas in tangible forms, while a trademark is a word, phrase, symbol, or design that identifies and distinguishes the source of the goods or services of one party from those of others. In simpler terms, copyright law protects your creative work, while trademark law protects your brand. Both are important to understand when creating signage. When designing signage, it is important to ensure that your design does not infringe on any existing copyrights or trademarks. This means avoiding using any images, logos, or phrases that may be protected by copyright or trademark. It is always a good idea to do thorough research and consult with a legal professional to ensure that your design is original and does not violate any laws.

Liability

As a signage designer, you are responsible for ensuring that your design is not only visually appealing, but also safe and functional. This means considering potential hazards and adhering to applicable safety regulations. For example, if your signage is placed outdoors, it must be able to withstand weather conditions and not pose a hazard to pedestrians. In addition, if your design includes any electronic components, it is important to follow safety guidelines to prevent any electrical issues.

It is also important to consider liability when it comes to the installation and maintenance of your signage. If any accidents or injuries occur due to faulty installation or poor maintenance, you could be held liable. Be sure to work closely with the installation and maintenance team to ensure that all safety protocols are followed.

Intellectual Property

As a designer, you are constantly creating new and innovative designs. It is important to protect your intellectual property and ensure that your design is not stolen or replicated without your consent. This is where trademarks and copyrights come into play again. By registering your design with the appropriate governing bodies, you can have legal protection against anyone reproducing your design without your permission.

In addition, it is important to have a clear understanding of ownership when it comes to design projects. Make sure that any contracts or agreements clearly outline who owns the design and the rights to use it. This will protect you and your work from potential intellectual property disputes in the future.

Wrapping Up

In conclusion, signage design goes beyond creating a visually appealing design. It is important to understand the legal regulations and considerations that come along with creating and installing signage. By being aware of copyright and trademark laws, considering liability, and protecting your intellectual property, you can ensure that your design is not only beautiful, but also legally sound.

Chapter 30: The Future of Signage and Wayfinding Design

The world of signage and wayfinding design is constantly evolving, with new technologies, emerging trends, and sustainable practices shaping the way we navigate and communicate in physical spaces. As designers, it is important to stay up-to-date and anticipate the future of our field in order to create effective and impactful signage solutions. In this chapter, we will explore the exciting innovations and possibilities on the horizon for signage and wayfinding design.

Technological Advancements

Technology continues to revolutionize the way we interact with the world around us, and signage is no exception. With advancements in digital displays, augmented reality, and artificial intelligence, the possibilities for signage and wayfinding design are expanding exponentially. Augmented reality has the potential to transform the way we navigate through physical spaces. By overlaying digital information on real-world environments, signage can become more interactive and dynamic, providing users with real-time information and personalized navigation. Artificial intelligence (AI) is also making its way into signage design, with the ability to analyze user behavior and adjust signage accordingly. For example, AI can track how people move through a space and use that data to improve wayfinding and signage placement for better user experience.

Emerging Trends

As the world becomes more digitally connected, it is not surprising that there is a trend towards incorporating technology into physical signage. However, there are also emerging trends that focus on the human element of signage and wayfinding design. One example is the rise of inclusive design, where signage is designed to be accessible and understandable for people of all abilities. This includes considerations for different cultures, languages, and special populations. Inclusive design not only improves user experience, but it also promotes a more inclusive and welcoming environment.

Another trend is the use of multidimensional design. With the advancement of materials and fabrication techniques, signage is no longer limited to traditional 2D forms. Designers are now exploring new shapes, textures, and lighting techniques to create visually dynamic and engaging signage.

Sustainable Practices

In recent years, there has been a growing awareness and concern for the environment, and this has extended to signage and wayfinding design as well. Sustainable practices are becoming a key consideration in the design process, from the materials used to energy consumption and maintenance. The use of environmentally-friendly materials, such as recycled plastics and sustainably sourced wood, is becoming more common in signage design. Additionally, incorporating renewable energy sources, such as solar power, into signage systems can reduce the overall carbon footprint of a space.

Maintenance and updating strategies are also being reevaluated through a sustainability lens. By choosing materials and techniques that require less frequent replacements or updates, the environmental impact of signage can be reduced.

Looking Towards the Future

As technology continues to advance at a rapid pace, it is important for designers to look towards the future and adapt our practices accordingly. However, while we embrace technological advancements and emerging trends, it is crucial that we do not lose sight of the human element in signage and wayfinding design. By incorporating sustainable practices, we can create a more responsible and inclusive future for the use of signage in physical spaces.

As we move forward, collaboration between designers, engineers, manufacturers, and users will be essential to push the boundaries of what is possible in signage and wayfinding design. By working together, we can create innovative and effective solutions that enhance the user experience and make navigating through physical spaces a seamless and enjoyable journey. This, ultimately, is the future of signage and wayfinding design – a harmonious blend of technology, inclusivity, and sustainability. But with endless possibilities and new developments on the horizon, the sky's the limit for this dynamic and ever-evolving field.

Chapter 31: Successful Signage and Wayfinding Design Examples

Signage and wayfinding design play a crucial role in shaping the user experience and in communicating important information. However, not all signage and wayfinding strategies are created equal. In this chapter, we will explore some successful examples of signage and wayfinding design and the valuable lessons they can teach us. From incorporating user-centered approaches to embracing innovative technology, these examples showcase the endless possibilities in creating effective and impactful signage and wayfinding solutions for various environments.

Lessons Learned

One of the key lessons we can learn from successful signage and wayfinding design examples is the importance of understanding the user. A user-centered approach ensures that the design is tailored to the specific needs, preferences, and behaviors of the target audience. This can be seen in the signage and wayfinding design for the Singapore Metro system. The design utilizes symbols and colors that are easily recognizable for both tourists and locals, making navigation effortless and enjoyable for all users. Incorporating elements of surprise and delight can also greatly enhance the user experience. This can be seen in the design of the Los Angeles Zoo's wayfinding system. The use of playful animal illustrations and witty signage not only helps visitors navigate the zoo but also adds a touch of charm and whimsy to their overall experience.

Another important lesson is the power of simplicity in design. Too much information and cluttered designs can overwhelm and confuse users. The signage and wayfinding design for Central Park in New York City is a great example of how simplicity can be highly effective. The use of minimalistic design and clear navigation cues allow visitors to easily find their way around the park without being bombarded with unnecessary information.

Design Elements to Emulate

There are certain design elements that are consistently present in successful signage and wayfinding strategies. These elements can serve as inspiration for designers looking to create impactful solutions. Effective use of typography is one such element. Typography not only adds visual interest but also plays a crucial role in guiding users. The signage and wayfinding design for the Philadelphia Museum of Art effectively utilizes typography to create a cohesive and aesthetically pleasing experience for visitors. Strategic placement of signage is another key element. Signage should not only be placed in visible and accessible locations but also guide users in the right direction. The wayfinding design for Tokyo Midtown incorporates both elements by using floor signs and wall signs to direct visitors to various destinations within the building. Furthermore, incorporating branding elements in signage and wayfinding design can greatly enhance the overall user experience. The signage for the Disney theme parks is a prime example of this. The use of familiar Disney characters, fonts, and colors not only aids in wayfinding but also adds to the magical and immersive experience of these iconic attractions.

Celebrating Artists and Cultures through Design

Signage and wayfinding design can also serve as a platform for showcasing local culture and promoting artists. The Dubai International Airport's wayfinding design does just that by incorporating traditional Arabic art and calligraphy into the design. This not only adds a unique aesthetic but also celebrates the local culture and adds a sense of place to the airport. Another example is the signage and wayfinding design for the University of the Arts London. The design incorporates the works of various alumni artists, creating an engaging and ever-changing experience for students and visitors.

By embracing different art forms and cultural elements in signage and wayfinding design, we can elevate the user experience and create a sense of connection and inclusivity within the space.

Embracing Technology

In today's digital age, incorporating technology into signage and wayfinding design can greatly enhance the user experience and increase efficiency. The London Gatwick

Airport's wayfinding system is a great example of this. By utilizing interactive touch screens and real-time information, the airport has reduced confusion and improved the overall travel experience for passengers. Similarly, the use of augmented reality in signage and wayfinding design has also shown great potential in creating immersive and engaging experiences. A great example of this is the AR wayfinding system at the Audi Forum in Ingolstadt, Germany, where users can simply point their smartphone camera at the signage to receive real-time directions.

Conclusion

The successful signage and wayfinding design examples discussed in this chapter showcase the endless possibilities in creating effective and impactful solutions. By understanding and incorporating elements such as user-centered design, simplicity, and strategic use of technology, we can enhance the user experience and create meaningful and memorable spaces for all. Let these examples inspire your next signage and wayfinding design project to create a truly exceptional experience for all users.

Chapter 32: Signage in Crisis Situations

When we think of signage and wayfinding design, our minds often jump to the mundane and practical aspects of everyday life. However, in times of crisis, these seemingly ordinary design elements take on a whole new level of importance. Natural disasters, military operations, and emergency response situations all require clear and effective signage to aid in communication, navigation, and safety. In this chapter, we will delve into the critical role of signage in crisis situations and explore how design can help support and protect individuals and communities.

Natural Disasters

Natural disasters, whether they be hurricanes, floods, wildfires, or earthquakes, are unfortunate events that can strike unexpectedly and with devastating consequences. In these situations, having clear and concise signage can be a matter of life and death. From evacuation routes to emergency shelters, effective signage is crucial in helping people find their way to safety. In addition to providing crucial information, signage can also offer a sense of reassurance and comfort in chaotic and frightening situations. The use of simple design elements, such as calming colors and familiar symbols, can help alleviate anxiety and promote a sense of security for those who may be feeling overwhelmed. During a natural disaster, it is essential that signage be durable and weather-resistant. Materials such as vinyl and aluminum are often used for their durability and ability to withstand harsh conditions. Signage should also be placed strategically in areas where it is visible and easily identifiable, even in low light or inclement weather.

Military Operations

In military operations, clear and concise communication is vital not only for the safety of military personnel but also for the success of the mission at hand. In these high-stress situations, effective signage can be a critical tool in relaying instructions, warnings, and other important information. In addition to providing critical information, signage can also play a role in maintaining secrecy and security. Military facilities often use coded or symbol-based signage to ensure that only those with the necessary clearance can access restricted areas. This type of signage requires specialized design skills and a deep understanding of military protocols and operations. Furthermore, in hostile and foreign environments, signage can serve as a universal language and aid in communication between military personnel and local populations. This is especially important in situations where there may be a language barrier, allowing for a smoother and more effective operation.

Emergency Response

In emergency response situations, time is of the essence, and effective communication can make all the difference. Whether it be a fire, a medical emergency, or a terrorist attack, clear and concise signage can help emergency responders quickly locate and assist those in need. Emergency response signage not only provides critical directions but also serves as a wayfinding tool for responders trying to navigate unfamiliar and chaotic environments. In situations where there may be multiple responding agencies, clear and standardized signage can promote cooperation and coordination. Designing for emergency response situations requires a deep understanding of emergency protocols and procedures. Signage must be designed to be easily visible and identifiable, even in highly stressful and crowded situations. This may include the use of high-contrast colors, bold font, and universal symbols that can be quickly understood by all.

In Conclusion

In times of crisis, clear and effective signage can be a lifeline for individuals and communities. From aiding in evacuation to supporting military operations and emergency response, signage plays a crucial role in promoting safety and saving lives. With careful design and consideration, signage can be a powerful tool in helping individuals and communities navigate through even the most challenging situations.

Chapter 33: Signage and Community Involvement

Participatory Design: Engaging the Community in Signage Design

Signage design is not just about visual aesthetics and navigation, it is also about creating a sense of community and belonging. In recent years, there has been a shift towards participatory design, where community involvement is seen as an essential part of the design process. This approach allows for a more inclusive and culturally sensitive design, as it takes into consideration the needs and preferences of the people who will be using the space. By involving the community in the design process, signage can become a powerful tool for bringing people together and creating a sense of ownership over their environment. Participatory design is rooted in the idea of co-creation, where designers work collaboratively with community members to develop a shared vision and design solution. It goes beyond the traditional top-down approach of design, where experts dictate the final outcome. Instead, it embraces the diversity of perspectives and knowledge within a community, allowing for a more holistic and meaningful design.

In the context of signage design, participatory design involves engaging the community in activities such as surveys, workshops, and focus groups to gather insights and feedback. These interactions not only help to generate data and ideas but also foster a sense of ownership and pride in the community. By involving community members in the design process, they feel valued and invested in the outcome, making them more likely to engage with and respect the signage.

Community Involvement: A Collaborative Effort

Community involvement in signage design goes beyond just seeking feedback; it is about creating a collaborative effort between designers and community members. This approach recognizes that the community has a deep understanding of the local area and its cultural context. By working together, designers can tap into this local knowledge and create signage that is relevant, culturally sensitive, and engaging. Involving the community in the design process can also bring about a sense of pride in their surroundings. By creating signage that reflects their culture and values, community members can feel a stronger connection to their environment. This can have a positive impact on their well-being and sense of belonging. Furthermore, community involvement allows for the identification and addressing of potential issues or concerns before the signage is implemented. This can help to avoid costly mistakes and ensure that the signage is effective in meeting the needs of the community. By working together, designers and community members can find solutions that benefit everyone.

Public Feedback: The Importance of Listening to the Community

In addition to involving the community in the design process, it is crucial to seek and listen to public feedback throughout the different stages of signage design. This feedback can come from various sources, such as surveys, public hearings, and online platforms. By actively seeking feedback, designers can evaluate the effectiveness of their design and make necessary adjustments to ensure it meets the needs of the community. Public feedback can also provide valuable insights into the community's perception and interpretation of the signage. It can reveal potential cultural sensitivities or misunderstandings that may arise, allowing for adjustments to be made to the design before installation. This not only avoids potential conflicts but also ensures that the signage is respectful and inclusive.

Finally, public feedback can help to create a sense of transparency and accountability in the design process. By involving the community and considering their feedback, designers are showing that they value their input and are committed to creating a design that truly serves the community.

In Conclusion

Involving the community in signage design not only leads to more meaningful and inclusive designs, but it also fosters a sense of community and pride in their environment. By embracing the diverse perspectives and local knowledge of community members, designers can create signage that is culturally sensitive, relevant, and engaging. Additionally, seeking and listening to public feedback ensures that the signage is effective, respectful, and reflects the community's values. Ultimately, community involvement in signage design is a collaborative effort that can have a positive impact on both the design process and the community's well-being.

Chapter 34: Signage and User Perception

Signage and wayfinding design are crucial elements in creating a positive and efficient user experience. However, the success of these designs relies heavily on the perception of users. In this chapter, we will explore the importance of user feedback, satisfaction surveys, and how perception can sometimes differ from reality in the world of signage and wayfinding design.

User Feedback

User feedback is an essential aspect of any design process and is especially important in the world of signage and wayfinding. It allows designers to gather information directly from the users, which can be used to improve and enhance the overall design. Feedback can be collected in various forms, such as face-to-face interviews, surveys, or through online platforms. In the context of signage and wayfinding design, user feedback can provide valuable insights into how users interact with the signage and what improvements can be made. For example, if users consistently have trouble navigating a certain area despite clear signage, it may indicate a need for a different approach or a redesign of the signage. Additionally, user feedback can also help designers understand the different needs and perspectives of diverse user groups, taking into consideration factors such as age, culture, and abilities.

Satisfaction Surveys

Satisfaction surveys are another helpful tool in understanding user perception and satisfaction with signage and wayfinding design. These surveys can be conducted periodically or at specific intervals to gather information about user experience and their level of satisfaction with the design. They can be administered through email, on-site kiosks, or through mobile apps. Satisfaction surveys not only provide valuable data for designers to improve their designs, but they also show that the company or organization values the opinions of their users. It creates a sense of inclusivity and promotes a customer-centric approach in the design process.

Perception vs. Reality

In the world of signage and wayfinding design, perception can often differ from reality. Signs and wayfinding elements may be perceived differently by different users, leading to confusion and frustration. Therefore, it is essential for designers to understand how users perceive and interpret signage. A common example of this is the use of symbols and icons in signage. While these symbols may seem universal, they can have different meanings in various cultures or even amongst different generations. Designers need to be mindful of these differences and ensure that the signage is understandable for all users.

Another aspect to consider is the environment in which the signage is placed. The lighting, background, and surrounding elements can all affect how users perceive and interpret the signage. Therefore, designers should conduct thorough research and on-site observations to ensure that the signage is effective in its intended environment.

In Conclusion

In conclusion, user feedback and satisfaction surveys are crucial tools in understanding user perception and satisfaction with signage and wayfinding design. They help designers to improve and enhance their designs, ultimately leading to a better user experience. As for the difference between perception and reality, designers must take into consideration various factors such as cultural differences, user preferences, and the environment to create effective and user-friendly signage. By continuously gathering feedback and conducting satisfaction surveys, designers can ensure that their signage meets the needs and expectations of their users.

Chapter 35: Designing for Large Spaces, Budgetary Constraints, and Changing User Needs

Large spaces, whether in public or private environments, present a unique challenge for designers when it comes to signage and wayfinding. These spaces require careful consideration when it comes to placement, design, and execution in order to effectively guide and enhance the user experience. Additionally, budgetary constraints and changing user needs must be taken into account in order to create a successful signage system. In this chapter, we will explore the key elements to consider when designing for large spaces, including budgetary constraints and changing user needs.

Designing for Large Spaces

Signage and wayfinding in large spaces, such as airports, shopping malls, or outdoor public areas, require a different approach than smaller environments. The sheer size of the space presents challenges in terms of finding the right balance between providing enough information without overwhelming the user. Here are some key considerations for designing signage in large spaces:

- Prioritize information: In large spaces, it is important to prioritize the most crucial information to prevent clutter and confusion. This means identifying the most important destinations, points of interest, and routes, and highlighting them through size, color, or placement.

- Use a consistent design: Consistency is key in creating an effective signage system in large spaces. The same visual language, colors, and typography should be used throughout the space to create a cohesive and unified experience for the users.

- Incorporate landmarks: In a vast and unfamiliar environment, landmarks can help users orient themselves and remember their path. Incorporating natural or designed landmarks, such as unique sculptures or eye-catching architectural features, can aid users in navigation and enhance the overall experience.

- Consider lighting and visibility: In large spaces, lighting and visibility can greatly impact the effectiveness of signage. Proper lighting can make a sign stand out and improve legibility, while low visibility can make it difficult for users to locate and read signs. Designers must take into account the lighting conditions of the space and choose materials and colors that will ensure visibility in all situations.

Budgetary Constraints

Designing for large spaces also means working within a budget, which can present its own set of challenges. In order to create an effective and visually appealing signage system, designers must find creative ways to work within these constraints. Here are some key strategies to consider:

- Simplify design: In order to save costs, designers may need to simplify the visual design of signage. This can mean using fewer colors, an alternate typeface, or reducing the size of signs while still maintaining readability.

- Use materials wisely: Choosing the right materials can have a significant impact on the cost of a signage system. Instead of opting for expensive materials, designers can find creative and cost-effective solutions, such as vinyl instead of hand-painted signage or digital screens instead of static signs.

- Partner with manufacturers: Collaboration with signage manufacturers can also help to reduce costs. These professionals have knowledge and experience with materials and production processes and can offer valuable suggestions for cost-saving measures.

Changing User Needs

As spaces and user needs evolve, signage systems must also adapt. It is important to regularly review and update signage design in order to meet the changing needs of users. Here are some key considerations for accommodating changing user needs:

- Conduct user research: Regularly conducting user research can help designers understand how users are utilizing the space and identify any areas where the signage system is not meeting their needs. This information can then be used to improve the system.

- Integrate technology: With the rise of technology, incorporating digital and interactive elements into signage systems can greatly enhance the user experience. Digital screens or interactive kiosks can provide real-time updates, directions, and other useful information.

- Design for inclusivity: Inclusivity should always be a priority in signage design, and this is especially important in large spaces. Designers must consider the needs of people with visual, hearing, or cognitive impairments when creating a signage system. This may include providing Braille, using high-contrast colors, or incorporating audio elements.

As designers continue to push the boundaries and create innovative solutions for signage and wayfinding in large spaces, careful consideration of design, budget, and user needs will be key to success. By incorporating these elements, designers can create effective and visually appealing signage systems that enhance the user experience and meet the unique challenges of large spaces.

Chapter 36: Collaborating for Successful Signage and Wayfinding Design

Communication, teamwork, and interdisciplinary collaboration are the cornerstones of successful signage and wayfinding design. In this chapter, we will explore the importance of these elements and how they contribute to creating effective and impactful signage that enhances the user's experience through semiotics.

Interdisciplinary Collaboration

Signage and wayfinding design is a multifaceted process that requires input and expertise from various disciplines. From graphic designers to urban planners, architecture to psychology, the collaboration of these diverse fields brings a richness and depth to the design process. In the world of signage and wayfinding design, none of these disciplines can stand alone. Each one plays a crucial role in creating a cohesive and functional system that meets the needs of the users. For example, a graphic designer brings the visual communication skills necessary to create legible and visually appealing signage, while an urban planner understands the layout of a space and how people move within it. When these disciplines work together, they can create a seamless and intuitive wayfinding system that benefits the users.

Teamwork

At the heart of interdisciplinary collaboration lies teamwork. In the context of signage and wayfinding design, teamwork is essential for a couple of reasons. First, it allows for a diversity of perspectives and ideas to be brought into the design process. This helps to avoid the pitfalls of tunnel vision and ensures that the needs and preferences of all stakeholders are considered. Secondly, effective teamwork allows for a more efficient and streamlined design process. With everyone working together towards a common goal, tasks can be delegated, and responsibilities can be shared. This not only saves time but also increases the quality of the end product as each team member can focus on their particular area of expertise.

Communication

Communication is the glue that holds interdisciplinary collaboration and teamwork together. Through effective communication, team members can express their ideas, provide feedback, and keep everyone on the same page. It is crucial to establish clear and open lines of communication from the beginning of the design process. It is also essential to communicate effectively with the clients and end-users. They are the ones who will ultimately be using the signage and wayfinding system, and their needs and preferences must be taken into account. By engaging in open and transparent communication with them, designers can ensure that the end product meets their expectations and enhances their experience.

Good communication also involves actively listening to others and being open to feedback and suggestions. This can help to avoid conflicts and misunderstandings and ensure a smoother design process.

The Benefits of Collaborating for Signage and Wayfinding Design

Collaboration in signage and wayfinding design brings numerous benefits, both for the team and the end-users. By working together and combining their expertise, designers can create effective and impactful signage that meets the needs of the users and enhances their experience. Collaboration also leads to increased innovation and creativity. When team members from different disciplines come together, they bring their unique perspectives and ideas, creating a melting pot of creativity that can result in new and innovative solutions. In addition, teamwork and collaboration can also lead to cost savings and efficiency. With everyone working together, tasks can be completed more efficiently, resulting in a faster and more streamlined design process. This can ultimately lead to cost savings for the client.

The Challenges of Collaborating in Signage and Wayfinding Design

While collaboration can bring many benefits, it also comes with its challenges. Effective communication and teamwork require effort, and it is not always easy to bring together team members from different backgrounds and disciplines. Additionally, conflicts and differences of opinions may arise, which can slow down the design process. However, by acknowledging these potential challenges and actively working to address them, designers can reap the many benefits of collaboration and teamwork.

Conclusion

Collaborating in signage and wayfinding design is crucial for creating effective and impactful signage that enhances the user's experience through semiotics. By fostering interdisciplinary collaboration, promoting teamwork, and open and effective communication, designers can create signage and wayfinding systems that not only meet the needs of the users but also communicate their messages and enhance their experience seamlessly.

Chapter 37: User-Centered Evaluation

User Surveys

Signage and wayfinding design cannot solely rely on the designer's perspective. It is essential to consider the opinions, preferences, and experiences of the users. User surveys are an effective way to gather insights from the target audience. It involves creating a set of questions and distributing it to a sample group of users, either physically or online. The results of these surveys can provide valuable information on how users interact with signage and their overall satisfaction with the design.

One of the key benefits of user surveys is that it allows designers to understand the needs and expectations of users. It also helps in identifying any issues or challenges that users may face while navigating through a space. By incorporating user feedback, designers can make improvements that enhance the user experience and make the design more effective.

Observation

Sometimes, actions speak louder than words. Observation is a powerful tool in evaluating the effectiveness of signage and wayfinding design. Designers can observe how users interact with the signage in real-time, allowing them to see firsthand how the design influences user behavior. It is essential to observe users from various backgrounds and abilities to gain a comprehensive understanding of the design's impact. Designers can also employ different observation techniques, such as recording videos or taking notes, to analyze the user's behavior and identify any patterns or issues. This method enables designers to make data-driven decisions and improve the design based on actual user behavior.

Eye Tracking

In recent years, eye tracking technology has become popular in the field of design and advertising. It involves measuring and recording eye movements to analyze the areas that capture a user's attention. Eye tracking can be a powerful tool in evaluating the effectiveness of signage and wayfinding design. By tracking eye movements, designers can determine whether users notice the signage and how long they spend reading or looking at it. This data can help designers understand how to make the design more visually appealing and prominent to the target audience. Eye tracking can also provide insights into any confusion or difficulty users may experience while navigating through a space. Incorporating eye tracking in the design process can significantly improve the user experience. By understanding what catches a user's attention, designers can make informed decisions on how to design the signage for better communication and navigation.

User-centered evaluation is a crucial aspect of signage and wayfinding design. By gathering feedback from users and observing their behavior, designers can continuously improve the design and make it more effective. It is also essential to involve users in the design process to ensure that their needs and preferences are considered. Ultimately, the goal of user-centered evaluation is to enhance the user experience and create a seamless and enjoyable journey for all.

Chapter 38: Ethical Considerations in Signage Design

Signage and wayfinding design serve a crucial role in guiding users and enhancing their experiences in various environments. However, as designers, it is our responsibility to not only create functional and aesthetically pleasing signage but also consider the ethical implications of our work. In this chapter, we will explore three important factors to consider when designing signage: cultural appropriation, stereotyping, and misinformation.

Cultural Appropriation

In recent years, cultural appropriation has been a hot topic in the design industry, and it is something that we must pay close attention to when designing signage. It refers to the adoption of elements from a culture that is not one's own, often without understanding or respecting their original meaning and significance. Cultural appropriation in signage design can be seen when symbols or imagery from a certain culture are used without proper context or understanding, leading to misrepresentation and disrespect.

To avoid cultural appropriation in signage design, it is essential to research and understand the cultural significance of the symbols and imagery used. It is also important to involve individuals from the respective culture in the design process and seek their feedback and approval. By doing so, we can create signage that is culturally sensitive and respectful.

Stereotyping

Stereotyping is another ethical consideration that must be addressed when designing signage. It refers to the generalization of certain characteristics or behaviors of a group of people based on their race, gender, religion, or other factors. Stereotypes can be harmful and offensive, and it is our responsibility as designers to avoid perpetuating them through our work.

One way to avoid stereotyping in signage design is by using diverse and inclusive visuals. This means representing a diverse range of people and cultures in the design, instead of relying on one dominant group. It is also important to be mindful of the language used in the signage, avoiding any terms or phrases that perpetuate stereotypes. By promoting diversity and inclusivity in signage, we can create a more welcoming and positive user experience.

Misinformation

Misinformation in signage design refers to the use of incorrect or misleading information. This can have severe consequences, especially in critical environments such as hospitals or public safety areas. As designers, we must ensure that the information presented on signage is accurate and consistent with the surrounding environment. To avoid misinformation in signage, it is crucial to fact-check all information and have it reviewed by experts if necessary. It is also important to consider the target audience when designing the signage and using language that is appropriate and easily understood by them. By taking these steps, we can create signage that not only looks appealing but also presents the correct information to users. As designers, it is our responsibility to create ethical signage that respects different cultures, avoids harmful stereotypes, and presents accurate information. By being mindful of these considerations, we can contribute to a more positive and inclusive society through our work.

In the next chapter, we will explore the concept of inclusive signage design and its importance in creating an accessible and welcoming environment for all users.

Chapter 39: Inclusive Signage Design

Design for All

Inclusive design, also known as universal design, is the concept of creating products, environments, and services that are usable by everyone, regardless of age, ability, or status. In the context of signage and wayfinding design, this means creating visuals and messaging that can be understood and utilized by all individuals, regardless of their physical or cognitive abilities. Designing for all is not only a moral responsibility but also a smart business decision, as it opens up the user experience to a wider audience and promotes inclusivity and diversity.

Universal Design Principles

The seven principles of universal design, developed by Ronald L. Mace, provide a framework for creating products and environments that are accessible to all individuals. These principles include flexibility in use, simple and intuitive use, equitable use, perceptible information, tolerance for error, low physical effort, and size and space for approach and use. Incorporating these principles into signage and wayfinding design can greatly enhance the user experience for individuals with disabilities, cognitive impairments, or language barriers.

Accessibility Audit

An accessibility audit is a comprehensive review of a space or product to determine its level of accessibility for individuals with disabilities. Conducting an audit for signage and wayfinding design involves evaluating the design elements, placement, and messaging to ensure they meet universal design principles and are accessible to everyone. This can include assessing the use of clear and concise language, high contrast colors, appropriate font sizes, and accessible placement of signage for individuals using mobility aids. In addition to physical accessibility, an accessibility audit also involves considering the needs of individuals with different disabilities. For example, individuals with visual impairments may require audible cues or tactile design elements, while individuals with cognitive impairments may benefit from simplified visuals and messaging. By conducting an accessibility audit throughout the design process, designers can ensure that their signage and wayfinding systems are inclusive and welcoming for all users. Designing for all should not be viewed as an added cost or burden, but rather as an essential aspect of good design. Not only does it promote inclusivity and diversity, but it also allows for a more enjoyable and efficient user experience for everyone. By incorporating universal design principles and conducting accessibility audits, signage and wayfinding design can truly enhance the user experience and create a more inclusive and welcoming environment. To ensure that signage and wayfinding systems are designed with all users in mind, it is important for designers to collaborate with individuals with disabilities during the design process. This can provide valuable insights and perspectives on how signage and wayfinding systems can be improved to better serve all users. Involving community engagement and user feedback is also crucial in creating a truly inclusive design.

In conclusion, creating signage and wayfinding systems that are accessible to all individuals is not only a moral responsibility, but also a smart business decision. By incorporating universal design principles and conducting accessibility audits, designers can enhance the user experience and promote inclusivity and diversity. Collaboration with individuals with disabilities and community engagement is key in designing for all and creating a world where everyone feels welcome and included. Let us continue to use our design skills to make the world a more inclusive and accessible place for all.

Chapter 40: Signage in the Time of COVID-19

Signage and wayfinding design has always been an essential aspect of the built environment, guiding and informing individuals in public spaces. However, with the emergence of the global COVID-19 pandemic, the role of signage has become even more critical. As we navigate this new normal, designers, architects, and urban planners are faced with the challenge of incorporating COVID-19 safety measures into their designs while maintaining their aesthetic and functional standards. In this chapter, we will explore the impact of the pandemic on signage and wayfinding design and the strategies being employed to mitigate its effects.

Social Distancing

One of the most significant factors influencing signage and wayfinding design in the wake of COVID-19 is the need for social distancing. As a result, physical markers, barriers, and directional arrows have become common features in public spaces, guiding people to maintain a safe distance from one another. Signage and wayfinding designers are tasked with finding creative ways to incorporate these measures into their designs without compromising their visual appeal. This has led to the use of innovative materials and designs for social distancing signage, such as tapestries with directional arrows, colorful decals on the ground, and three-dimensional markers. Designs that were once solely functional are now being elevated to serve as aesthetically pleasing features, reinforcing the concept of social distancing while adding to the overall ambiance of the space.

Health and Safety Signage

Aside from social distancing, health and safety signage has also become a critical aspect of signage design in the age of COVID-19. These signs serve as reminders to individuals to wash their hands, wear masks, and maintain proper hygiene practices. They also inform the public of the rules and regulations in place to ensure their safety.

However, health and safety signage must strike a balance between being informative and not causing panic or anxiety. Designers must carefully consider the tone and language used in these signs to deliver the necessary information while maintaining a sense of calm and reassurance.

Contactless Technology

The pandemic has also accelerated the adoption of contactless technology in the design of signage and wayfinding systems. With the fear of contamination through touch, designers are exploring ways to incorporate contactless technologies such as QR codes, voice activation, and gesture control in their designs. These technologies not only reduce the risk of transmission but also enhance the overall user experience by providing a touch-free and efficient means of navigation. Furthermore, with the widespread use of smartphones, designers are incorporating mobile apps and augmented reality technology into their signage designs. This not only serves as a contactless wayfinding solution but also provides a personalized experience for the user.

Collaboration in the Face of Crisis

The challenges posed by COVID-19 have emphasized the importance of collaboration in the design of signage and wayfinding systems. The role of designers, architects, and urban planners in creating safe and functional spaces has become even more critical. In response, there has been a push towards collaborative design processes, involving all stakeholders in the planning and implementation of COVID-19 safety measures.

Furthermore, designers are also collaborating with signage manufacturers, utilizing their expertise and resources to produce the necessary signage in a timely and efficient manner. This collaboration not only ensures the effective implementation of safety measures but also supports the local economy.

Adapting to a Changing World

The pandemic has brought about significant changes in the way we navigate and interact with public spaces. As such, designers must also be adaptable and open to change in their approach to signage and wayfinding design. The traditional wayfinding strategies may no longer be applicable in a post-COVID world, and it is essential for designers to stay informed and updated on the latest developments and guidelines.

In addition, designers must also consider the longevity of their designs and the possibility of future pandemics or crises. This has led to an emphasis on modular and flexible designs that can be easily adapted and updated in response to changing circumstances.

Conclusion

The COVID-19 pandemic has undoubtedly had a significant impact on signage and wayfinding design. However, as designers, we have the opportunity to use this as a catalyst for innovation and creativity. By incorporating COVID-19 safety measures into our designs while maintaining our standards of aesthetics and functionality, we can enhance the user experience while keeping the public safe. Collaboration, adaptability, and a forward-thinking approach will be crucial in navigating the ever-changing landscape of signage and wayfinding design in the time of COVID-19.

Chapter 41: Signage in Urban Planning

Signage plays a key role in the overall design and functionality of a city. Effective signage can create a sense of place, aid in wayfinding, and enhance the overall user experience. As urban planning continues to evolve, the importance of signage becomes even greater. In this chapter, we will explore the various ways in which signage contributes to the urban landscape and how it can be used to create a more harmonious and efficient cityscape.

Wayfinding in Cities

Wayfinding is the process of navigating through an environment, whether it be physical or digital. In cities, the use of signage is crucial in providing clear and concise directions for residents and visitors. As cities continue to grow and expand, the need for effective wayfinding becomes even more imperative. Signs not only guide people to their desired destination, but they also create a sense of security and ease for those navigating unfamiliar areas. In urban planning, wayfinding is a key consideration in the placement and design of signage. It is important for signs to be easily visible, readable, and located in strategic locations. This requires a thorough understanding of the city's layout and the potential points of confusion for users. Through careful planning and execution, cities can create a wayfinding system that is both functional and visually appealing.

Public Transportation

As cities become more densely populated, the need for efficient and accessible public transportation systems increases. Signage plays a crucial role in guiding people through these systems. From bus and train stations to airport terminals, signage helps users navigate through the various modes of transportation and find their way to their desired destination. In addition to wayfinding, signage in public transportation also serves as a source of information. Important details such as schedules, routes, and fare prices can all be easily communicated through well-designed signs. In this way, signage not only aids in navigation, but it also enhances the overall user experience by providing necessary information in a clear and concise manner.

Placemaking

Placemaking is the concept of designing public spaces in a way that encourages community engagement and enhances the overall experience of those interacting with the space. Signage plays a key role in this process by not only providing information, but also by adding to the aesthetic and cultural value of a space. In cities, placemaking through signage can take various forms. From wayfinding signs that use local landmarks to guide people, to decorative signs that showcase the city's history and culture, signage can truly transform a space and create a sense of identity for a city. By collaborating with local artists and incorporating elements of the city's unique character, signage can become more than just a functional tool, but also an artistic and cultural statement. In conclusion, signage in urban planning is a crucial aspect of creating a functional, visually appealing, and culturally rich cityscape. Through effective wayfinding, public transportation guidance, and placemaking, signage can enhance the overall user experience and contribute to the identity of a city. As cities continue to evolve and grow, the importance of thoughtful and strategic signage will only continue to increase. By understanding the role of signage in urban planning and incorporating it into the design process, we can create cities that are not only functional, but also beautiful and inviting for all who interact with them.

Thank you for reading Chapter 41 of our book on Signage and Wayfinding Design – Enhancing the User Experience through Semiotics. In the next chapter, we will explore the role of signage in retail spaces and its impact on the overall brand experience. Stay tuned for more insightful chapters on this fascinating topic.

Chapter 42: Signage in Retail Spaces

Retail stores are more than just places to buy goods. They are spaces designed to evoke emotions, create experiences, and ultimately, drive sales. The success of a retail store relies not only on the products it sells, but also on its store layout, visual merchandising, and understanding of consumer behavior. As signage and wayfinding designers, we have a crucial role in enhancing the overall retail experience. In this chapter, we will explore the importance of store layout, visual merchandising, and consumer behavior in creating an enticing and successful retail environment.

Store Layout

When a customer enters a retail store, the first thing they notice is the overall layout. The way the store is organized and designed can greatly impact a customer's shopping experience. An effective store layout should be intuitive, flow smoothly, and lead customers to the products they are looking for. As signage and wayfinding designers, we should pay close attention to the store layout and how our signage can enhance it. The layout of a retail store should be tailored to its specific target market and product offerings. For example, a store that sells luxury goods may have a more spacious and upscale layout, while a discount store may have a more cluttered and busy layout to convey its low prices. The layout should also take into consideration the flow of traffic within the store. Placing popular items towards the back of the store or creating designated "hot spots" can encourage customers to explore more of the store and possibly make additional purchases.

The use of signage can greatly enhance the store layout and help guide customers to different sections and products. Strategically placed signage can also create a sense of direction and flow within the store. For example, placing large, eye-catching signage above each aisle can help customers identify what products are available in that section.

Visual Merchandising

Visual merchandising, the art of creating attractive and appealing product displays,

plays a crucial role in retail stores. Not only does it showcase products, but it also helps to capture the attention and interest of customers. Effective visual merchandising can increase sales and create a memorable shopping experience. As signage and wayfinding designers, we can collaborate with visual merchandisers to create impactful displays. Our signage should complement the overall aesthetic and theme of the display. It should also provide important information, such as pricing or product features, without taking away from the overall appeal of the display. In today's retail landscape, visual merchandising is not limited to just physical displays. With the rise of technology, digital displays and interactive elements have become increasingly popular in retail stores. As signage designers, it is important to incorporate these elements into our designs to create a dynamic and engaging visual experience for customers.

Consumer Behavior

In order to create effective signage and wayfinding in retail spaces, it is crucial to understand consumer behavior. This includes understanding their shopping habits, preferences, and motivations. With this knowledge, we can create signage that not only guides customers, but also influences their purchasing decisions. One key aspect of consumer behavior is the power of suggestion. When a customer sees a sign or display, it can trigger a desire for a certain product or reinforce a decision to make a purchase. Placing signage strategically throughout the store can have a significant impact on sales. Another important aspect is the attention span of customers. In today's fast-paced world, customers have limited time and patience. This means that signage and wayfinding should be clear, concise, and easy to understand. Using images and icons, in addition to text, can help to grab the attention of customers and convey information quickly. In addition to understanding consumer behavior, it is also important to regularly evaluate and modify signage and wayfinding designs based on customer feedback. By adapting to changing consumer preferences and behaviors, we can create a more effective and enjoyable retail experience for customers.

In conclusion, as signage and wayfinding designers, we have an important role to play in enhancing the retail experience for customers. By considering store layout, visual merchandising, and consumer behavior, we can create impactful signage that guides and influences customers, ultimately leading to increased sales and a positive shopping experience. With our creativity and understanding of the retail space, we can truly elevate the retail environment through semiotics.

Chapter 43: Signage and User Experience

Impact on Emotions and Behavior

When it comes to signage and wayfinding design, the impact on emotions and behavior is often overlooked. However, the truth is that signage has a significant influence on how people feel and act in a given space. A well-designed and clear signage system can elicit positive emotions, such as comfort, trust, and enjoyment, while a confusing and poorly designed system can cause frustration, confusion, and even anxiety. A study conducted by the International Sign Association found that effective signage can increase customer satisfaction and brand loyalty, leading to repeat visits and increased revenue. This shows the power of well-designed signage in shaping user behavior. But how exactly does signage elicit these emotions and impact behavior? First, it is important to understand that our emotions and cognitive processes are deeply connected. In other words, our emotions can also influence our thoughts and actions. Therefore, when users encounter signage that is visually appealing and easy to understand, they are more likely to feel positive emotions, leading to a more pleasant and stress-free experience. On the other hand, signage that is confusing or irrelevant can create a negative emotional response, hindering the user's ability to navigate and potentially impacting their behavior in a negative way.

In addition, signage can also affect user behavior through its ability to guide and direct. A well-designed wayfinding system can not only help users navigate a space but also influence their movement and decision-making. For example, strategically placed directional signage can guide users towards a particular area or help them choose a specific route. This can be particularly effective in retail spaces, where well-placed signage can lead to increased sales and customer engagement.

Design Psychology

Design psychology, also known as environmental psychology, is the study of how the physical environment impacts our thoughts, feelings, and behaviors. Within the realm of signage and wayfinding design, incorporating principles of design psychology can greatly enhance the overall user experience. One key aspect of design psychology is the idea that people are visual creatures. Our eyes are constantly scanning our surroundings, subconsciously evaluating the space we are in. This is where signage comes into play. By using bold typography, bright colors, and eye-catching graphics, design psychology can be used to captivate and engage users, making their experience more enjoyable and memorable. Another aspect of design psychology is the concept of environmental cues. These are visual or auditory signals within a space that can provide important information and influence behavior. In signage design, effective use of environmental cues can help users navigate and interact with their surroundings. For example, color-coded signage can be used to differentiate between different areas or departments in a building, making it easier for users to find their desired destination.

By incorporating elements of design psychology into signage and wayfinding design, designers can create a more engaging and user-friendly experience for their audience.

User Engagement

User engagement refers to the level of involvement and interaction between a user and a product or service. In the context of signage and wayfinding design, user engagement is crucial in ensuring the effectiveness and success of a system. An engaging signage system should not only be visually appealing but also interactive and informative. The best way to engage users is to make them an active participant in the navigation process. This can be achieved through the use of digital signage, interactive kiosks, and QR codes, which allow users to access information or directions with just a few taps on a screen. By involving users in the process, they feel more in control and are more likely to have a positive experience. In addition, user engagement can also be achieved through creative and unique ways of conveying information. For example, incorporating humor or interactive elements into signage can make users more receptive to the message and more likely to remember it.

Overall, user engagement is essential in creating a memorable and effective signage system that enhances the user experience.

In Conclusion

Signage and wayfinding design go far beyond providing direction or information. By understanding their impact on emotions and behavior, incorporating principles of design psychology, and focusing on user engagement, designers can create a truly immersive and enjoyable experience for users. By carefully considering these aspects, signage can become a powerful tool in not only guiding but also influencing user behavior and leaving a positive impression in their minds.

Chapter 44: Brand Experience: Designing with Purpose

In today's fast-paced and digital world, it is more important than ever for brands to tell a powerful and compelling story. Brand storytelling is the art of crafting a narrative that connects with consumers on an emotional level and also reflects the values and beliefs of the brand. In essence, it is the foundation of a brand's experience and plays a crucial role in shaping customer loyalty.

Brand Storytelling: The Heart of Brand Experience

At its core, brand storytelling is about creating an emotional connection with customers. It goes beyond simply promoting a product or service and delves into the deeper layers of a brand's identity. By sharing stories that resonate with their audience, brands can foster a sense of community and build a sense of trust and loyalty. In the context of signage and wayfinding design, brand storytelling is essential in creating a cohesive experience for customers. Every element and design choice should align with the brand's story and contribute to the overall experience. For example, a brand rooted in nature and sustainability may choose to use natural materials and earth tones in their signage, while a brand focused on technology and innovation may opt for sleek and modern designs.

Brand Values: The Soul of Brand Experience

Brand values are the guiding principles that define a brand's purpose and drive its actions. They represent the fundamental beliefs and morals that a brand stands for and are crucial in shaping its reputation and reputation. Brands that are successful in creating a powerful brand experience are those that are true to their values and reflect them in all aspects of their business, including design.

When it comes to signage and wayfinding design, brands must ensure that their values are reflected in the message and visual elements. This not only adds authenticity to the brand but also helps customers connect and relate to the brand on a deeper level. For example, a brand that values inclusivity and diversity may choose to feature diverse faces and languages in their signage, showcasing their commitment to these values.

Customer Loyalty: The Result of a Seamless Brand Experience

The ultimate goal of brand experience is to create customer loyalty. When customers feel connected to a brand, resonate with its story and values, and have a positive experience, they are more likely to become loyal customers. They will choose the brand over competitors and even become brand ambassadors, spreading the brand's message and advocating for its products or services. Effective brand signage and wayfinding design play a crucial role in achieving customer loyalty. By creating a seamless and immersive brand experience, customers feel more connected and invested in the brand. They will also have a better overall experience, as they can easily navigate and find what they need. This positive encounter with the brand will further solidify their loyalty and make them more likely to return in the future.

In conclusion, creating a powerful and cohesive brand experience through storytelling, reflecting brand values, and fostering customer loyalty is essential for any brand to thrive in today's market. In the world of signage and wayfinding design, this becomes even more important as these elements are the first point of contact for customers and can make or break their experience. By designing with purpose and keeping the brand's story and values in mind, designers can contribute to creating a remarkable brand experience for customers.

Chapter 45: Signage Design for Multiple Languages

Translation: Bridging Language Barriers in Signage Design

One of the most important considerations in creating effective signage is ensuring that it can be understood by all those who come across it. In a diverse and globalized world, this means catering to a multitude of languages. Signage design for multiple languages requires a unique approach that goes beyond mere translation.

Translation involves converting written text from one language to another. However, when it comes to signage design, mere translation of words may not be enough to effectively convey the intended message. This is where the art of localization comes in.

Localization: Beyond Words

Localization is the process of adapting a product or service to a specific cultural and linguistic context. When it comes to signage design, this means going beyond the words displayed and taking cultural nuances into consideration. This includes factors such as colors, symbols, and images that may hold different meanings in different cultures.

For example, the color red may signify danger in Western cultures, but it symbolizes good luck and happiness in Chinese culture. Ignoring these cultural differences can lead to confusion or even offense among users of the signage. Therefore, it is crucial to involve cultural experts in the design process to ensure that the signage is culturally appropriate and effectively conveys its intended message to all users.

Cultural Adaptation: Designing for Different Cultures

Cultural adaptation is the process of modifying a product or service to fit the preferences and expectations of different cultures. In signage design, this means considering the cultural background and behaviors of the targeted audience. For example, in some cultures, it is common for people to read from right to left instead of left to right. In these cases, signage should be designed with this in mind, to ensure that the message is read in the correct order. Similarly, cultural perceptions of time and space must also be taken into account. In some cultures, punctuality and efficiency are highly valued, while in others, a more relaxed approach is preferred. This can affect the design and placement of signage in various environments. Furthermore, cultural adaptation should also be applied to the use of symbols and imagery in signage. What may be familiar and easily understood in one culture may be completely foreign in another. For example, a picture of a hand pointing may be universally understood as a directional symbol, but in some cultures, pointing is considered rude and offensive. This is why it is essential to research and understand the cultural context in which the signage will be displayed to avoid any unintended miscommunications.

Creating a Harmonious Multilingual Design

Designing signage for multiple languages involves finding a balance between consistency and diversity. It is essential to maintain a consistent design aesthetic to ensure cohesion and brand identity. However, it is equally important to incorporate cultural diversity to cater to different languages and cultures. One way to achieve this is through the use of universal symbols and images that can be easily understood by people from different cultural backgrounds. Another approach is to use a simple and minimalist design that can be easily adapted for different languages. This allows for a harmonious and cohesive design while still accommodating the needs of a diverse audience.

Simplifying the Translation Process

In the past, creating multilingual signage involved a complex and time-consuming translation process. However, with advancements in technology and design, this process has become much simpler and more efficient. With the help of language translation software and design tools, translations can now be done quickly and accurately. This not only makes the process more efficient but also ensures consistency in the translated messages. Additionally, cloud-based translation platforms allow for real-time collaboration and feedback from linguists and cultural experts, further improving the accuracy and appropriateness of the signage design.

Cross-Cultural Collaboration for Effective Signage Design

Designing signage for multiple languages requires a collaborative approach between linguists, designers, and cultural experts. By involving all stakeholders in the process, the design can be tailored to meet the needs and expectations of a diverse audience. Additionally, this collaboration can provide valuable insights and feedback that may not have been considered otherwise. By embracing cross-cultural collaboration, signage designers can create designs that not only effectively convey their intended message but also resonate with the cultural backgrounds of their audience.

In Conclusion

Signage design for multiple languages goes beyond a simple translation of words. It requires a deep understanding of cultural nuances and an inclusive approach to design. Through the use of localization, cultural adaptation, and cross-cultural collaboration, signage designers can create designs that are both visually appealing and effectively convey their message to a diverse audience. As our world becomes increasingly globalized, it is essential for signage designers to embrace this approach in order to promote inclusivity and enhance the user experience for all.

Chapter 46: Typography in Signage Design

In the world of signage design, typography plays a crucial role in creating effective visual communication. The use of various typefaces can significantly impact the legibility, readability, and overall aesthetic appeal of a sign. In this chapter, we will delve deeper into the importance of typography in signage design and the factors that should be considered when selecting typefaces.

Legibility

When designing signage, legibility should be one of the top priorities. After all, if a sign is not easily readable, its message will not be effectively communicated. Legibility refers to how easily a typeface can be read, especially from a distance. The primary purpose of signage is to convey information quickly and efficiently, therefore, selecting a legible typeface is essential. There are a few factors to consider when determining the legibility of a typeface. The first is the size of the letters. In general, the smaller the size of the letters, the less legible the typeface will be. This is especially important to keep in mind when designing signage for outdoor spaces where the signs may be viewed from a distance. The spacing between letters, known as kerning, also plays a role in legibility. Proper kerning ensures that all letters are evenly spaced, making it easier for the reader to identify individual letters and words. Contrast is another crucial factor in legibility. A high contrast between the letters and the background helps to make the text stand out and easier to read. This is why often, black or dark-colored text is used on a white or light-colored background. Highly decorative typefaces may decrease the contrast, making them more difficult to read.

Typeface Selection

Selecting the right typeface for a particular sign is critical to its effectiveness. Different typefaces evoke different emotions and have their unique personalities. The typeface used for a business sign should accurately represent the company's brand and overall message. For example, a bold and clean sans-serif font may be more appropriate for a modern and minimalistic brand, while a more elegant serif font may better suit a high-end luxury brand.

Additionally, the context and purpose of the sign must also be considered when choosing a typeface. For example, a sign for a children's playground could use a fun and playful typeface, while a sign for a hospital would require a more professional and serious typeface. The legibility of the typeface should also be a factor in the selection process.

Communication Hierarchy

The use of different typefaces and typography techniques can help create a hierarchy of the information on a sign. This is beneficial in cases where multiple pieces of information need to be conveyed in a concise and organized manner. The most important information should be emphasized with a larger or bolder typeface, while secondary information can be presented in a smaller font. This helps the reader to quickly identify and understand the most critical information on the sign. Another way to establish a communication hierarchy is through the use of different font styles, such as italic or bold. These variations help to draw attention to specific words or phrases, creating a visual hierarchy of information. However, it is crucial to use these techniques sparingly to avoid overwhelming the reader and making the sign confusing.

In conclusion, typography is an essential aspect of signage design that should not be overlooked. The legibility, typeface selection, and communication hierarchy can significantly impact the effectiveness of a sign. By carefully considering these factors and utilizing typography techniques, designers can create visually appealing and easily readable signs that effectively communicate their message.

Chapter 47: Signage Design in the Digital Age

Welcome to Chapter 47 of our book on Signage and Wayfinding Design. As we delve into the world of modern visual communication, we will explore the exciting and ever-evolving field of digital signage design. With the rise of technology and its integration into our daily lives, the use of motion graphics, animation, and interactive interfaces has become increasingly popular in signage design. In this chapter, we will discuss the impact and possibilities of utilizing these elements in signage design, and how they can enhance the user experience.

Motion Graphics

Motion graphics refer to animated or moving visual elements that are designed to capture the viewer's attention and convey a message. In signage design, motion graphics can be used to create dynamic and eye-catching displays that are more likely to grab the attention of users. The use of motion can also make the messaging more memorable, as our brains process and retain visual information better than static images or text. One of the key advantages of using motion graphics in signage design is the ability to deliver a message in a short period of time. With our fast-paced lives, people are constantly on the move and have less time to consume information. Motion graphics can quickly convey important information in a visually appealing manner, making them highly effective in high-traffic areas such as train stations, airports, and shopping centers. Additionally, the use of motion graphics allows for a more immersive and interactive experience for the user. By incorporating elements such as movement, sound, and touch, designers can create a multi-sensory experience that engages the user on a deeper level. This can be particularly useful in environments where there is a need to communicate complex information or in engaging children and young adults.

Animation

Similar to motion graphics, animation is another element that can add a dynamic and engaging touch to signage design. By using sequential frames or images, animations create the illusion of movement, providing a playful and cheerful aesthetic to the signage. Animations can also be used to guide the user's attention and help them navigate through a space. By highlighting certain elements or using animated arrows and icons, designers can direct the user's gaze and guide them towards their intended destination. Moreover, animations can add an element of storytelling to signage design. By using characters, scenarios, and visual narratives, designers can create a more personal and emotional connection with the user. This can be particularly effective in public spaces such as museums, where stories and narratives play an important role in conveying information.

Interactive Interfaces

Interactive interfaces have revolutionized the way we interact with signs. With the use of touchscreens, sensors, and other interactive technology, designers can create a user experience that is not only informative but also entertaining and engaging. One of the key advantages of interactive interfaces is the flexibility it allows in terms of content and updates. Unlike traditional signage, updates and changes can be made easily and quickly, saving time and resources. This can be particularly useful in environments that require frequent updates, such as airports and shopping centers. Interactive interfaces also provide a sense of control and personalization to the user. By allowing them to actively engage with the signage, users feel more connected to the information being presented. This can also be beneficial for individuals with specific needs or preferences, as they can customize the content according to their needs.

In conclusion, the use of motion graphics, animation, and interactive interfaces in signage design has opened up a whole new world of possibilities. With the advancement of technology, the possibilities for creativity and innovation in signage design are endless. From creating jaw-dropping displays to providing a more personalized and engaging experience for the user, digital signage has become an integral part of modern visual communication. We hope this chapter has inspired you to explore and integrate these elements in your own signage designs. Thank you for reading and stay tuned for the next chapter on sustainability in signage design.

Chapter 48: Signage Design and Sustainability

Signage and wayfinding design have a significant impact on the environment, and as designers, it is our responsibility to incorporate sustainability into our work. In this chapter, we will explore how sustainable materials, energy-efficient design, and environmental impact assessments can be incorporated into signage design to create a more environmentally conscious and responsible approach.

Sustainable Materials

When it comes to signage design, there are various materials to choose from, each with its own pros and cons. However, as designers, we must prioritize using sustainable materials to minimize the environmental impact. This means using materials that have a lower carbon footprint, are renewable, and have a longer lifespan. One example of sustainable material is bamboo, which is a rapidly renewable resource that grows faster than traditional trees. Bamboo can be used for signage panels, frames, and even as a substrate for printing. With proper treatment and maintenance, bamboo can last for years and is a more environmentally friendly alternative to wood. Recycled materials such as aluminum and plastic can also be used in signage design. These materials not only reduce waste but also have a lower impact on the environment than producing new materials from scratch. By using recycled materials, we can save energy and reduce the emission of greenhouse gases.

Energy-Efficient Design

Incorporating energy-efficient design into signage can also help reduce the environmental impact. This includes using LED lights for illuminated signage, which are more energy-efficient and have a longer lifespan compared to traditional fluorescent or incandescent lights. LED lights also produce less heat, reducing the load on air-conditioning systems and further contributing to energy savings. Another aspect of energy-efficient design is the use of solar-powered signage. With advancements in technology, solar panels can now be integrated into signage structures, providing a sustainable and renewable source of energy. This not only reduces the carbon footprint but also lowers the operating costs of the signage.

Environmental Impact Assessment

Before embarking on any signage project, it is essential to conduct an environmental impact assessment to identify potential environmental impacts and develop strategies to address them. This assessment should consider factors such as the materials used, energy consumption, waste management, and the overall lifespan of the signage. By conducting an environmental impact assessment, we can identify and mitigate any potential negative impacts and ensure that the signage design is aligned with sustainability goals. This assessment should be an ongoing process throughout the project, from design to installation and maintenance.

In addition to conducting environmental impact assessments, it is also essential to involve stakeholders, such as local communities and environmental experts, in the design process. This collaborative approach can help identify any potential issues and develop solutions that benefit both the environment and the community.

Incorporating Sustainability in Signage Design

The incorporation of sustainable materials, energy-efficient design, and environmental impact assessments into signage design is not only beneficial to the environment but also to the overall user experience. Using sustainable materials can create a positive and lasting impression on users, who are becoming increasingly environmentally conscious and expect the same from the businesses they interact with. Moreover, adopting sustainable practices in signage design can also result in cost savings in the long run. For example, energy-efficient LED lights may have a higher upfront cost, but they last longer and consume less energy, resulting in cost savings over time.

The Future of Sustainable Signage Design

As society becomes more environmentally aware, the demand for sustainable products and services increases. This means that the future of signage design must prioritize sustainability, and designers must continue to find innovative ways to incorporate sustainability into their work. Technology, such as 3D printing, can also play a significant role in sustainable signage design. By using biodegradable or recycled materials in 3D printing, we can further reduce the environmental impact of traditional signage production methods. Furthermore, the continued collaboration between designers, manufacturers, and stakeholders will also be crucial in the development of sustainable signage design. By working together, we can create a more sustainable and responsible approach to signage design.

Conclusion

In conclusion, sustainability should be at the forefront of signage design to minimize the environmental impact and meet the expectations of environmentally conscious users. By using sustainable materials, incorporating energy-efficient design, and conducting environmental impact assessments, we can create signage that not only enhances the user experience but also aligns with our responsibility towards the environment. As designers, it is our duty to prioritize sustainability in our work and continue to find ways to improve and innovate for a more sustainable future.

Chapter 49: Collaborating with Signage Manufacturers

Selecting a manufacturer to produce your signage is an important decision that can greatly affect the overall success of your project. With the wide range of printing techniques, materials, and quality control processes available, it is crucial to carefully consider these factors in collaboration with your manufacturer to ensure the best possible outcome for your signage. In this chapter, we will explore these important aspects of working with signage manufacturers.

Printing Techniques

With the advancements in technology, there are now a variety of printing techniques available for creating high-quality signage. Each method offers its own distinct advantages, depending on the project's needs and requirements. Some of the most common printing techniques used by manufacturers include digital printing, screen printing, and laser engraving. Digital printing, also known as inkjet printing, is a widely used technique that uses specialized printers to transfer digital images directly onto various materials. The benefits of digital printing include high color accuracy, sharp images, and flexibility in producing different sizes and quantities of signage quickly. Screen printing is a traditional printing method that involves using a stencil to transfer ink onto a surface. This technique is ideal for creating bold and vibrant designs, making it a popular choice for outdoor signage. It also offers the advantage of being cost-effective for larger quantities of signage.

Laser engraving is a precise and efficient technique that involves using a laser to etch designs onto materials. This method is best suited for creating detailed and intricate designs, and it offers long-lasting results.

Material Selection

Choosing the right materials for your signage is crucial in ensuring its durability and visual impact. Your manufacturer can provide expert guidance on the best materials for your specific project. Some of the commonly used materials in signage production include acrylic, PVC, metal, and wood. Acrylic is a versatile material that offers a sleek and modern look. It is highly durable and can withstand harsh weather conditions, making it ideal for outdoor signage. Additionally, acrylic can be easily cut, shaped, and engraved, allowing for a wide range of design possibilities. PVC, or polyvinyl chloride, is a plastic material commonly used in signage production due to its cost-effectiveness and durability. It can be easily cut, painted, and printed on, making it a versatile option for various types of signage. Metal, such as aluminum and stainless steel, offers a professional and sophisticated look. It is a popular choice for indoor and outdoor signage due to its durability and aesthetic appeal. With metal, options for customization, such as embossing and engraving, are also available. Wood is a timeless material that exudes warmth and elegance. It offers a natural and rustic look, making it a popular choice for wayfinding signage and indoor environments. Wood signage can also be customized through various techniques such as carving and painting.

Quality Control

Collaboration with your signage manufacturer requires open communication and a mutual understanding of quality control standards. Quality control is essential in ensuring that the finished product meets your expectations and is delivered within the agreed timeline. Your manufacturer should have a reliable quality control process in place, which includes regular inspections, testing, and adherence to industry standards. It is also essential to establish clear communication channels for feedback and addressing any issues that may arise during production. As a client, it is crucial to be involved in the quality control process, whether it be through regular site visits or reviewing samples of the signage throughout production. This will not only ensure the quality of your signage but also provide an opportunity for any necessary modifications to be made before the final product is delivered.

Collaborating with your signage manufacturer can be a fruitful and rewarding experience if done successfully. By understanding the various printing techniques, materials, and quality control processes available, you can work closely with your manufacturer to bring your signage visions to life. Remember, communication is key, and with a collaborative approach, you can achieve the desired results for your signage project.

Printed in Great Britain
by Amazon